AGRICULTURAL TRADE LIBERALIZATION AND THE LEAST DEVELOPED COUNTRIES

Wageningen UR Frontis Series

VOLUME 19

Series editor:

R.J. Bogers
Frontis – Wageningen International Nucleus for Strategic Expertise,
Wageningen University and Research Centre, The Netherlands

Online version at http://library.wur.nl/frontis/

The titles published in this series are listed at the end of this volume.

AGRICULTURAL TRADE LIBERALIZATION AND THE LEAST DEVELOPED COUNTRIES

Edited by

NIEK KONING
Agricultural Economics and Rural Policy Group
Wageningen University and Research Centre, Wageningen, The Netherlands

and

PER PINSTRUP-ANDERSEN
Cornell University, Ithaca, NY, USA

A C.I.P. Catalogue record for this book is available from the Library of Congress.

ISBN 978-1-4020-6079-3 (HB)
ISBN 978-1-4020-6085-4 (PB)

Published by Springer,
P.O. Box 17, 3300 AA Dordrecht, The Netherlands.

www.springer.com

Front cover illustration: Henk van Ruitenbeek

Printed on acid-free paper

CONTENTS

FOREWORD

Economic welfare in many developing countries, and especially the least developed ones, has not increased during the last decades. This disconcerting fact has prompted an intensive debate on how this situation could be improved. The world community has formulated a number of Millennium Development Goals and has promised to adjust its policies to allow these goals to be realized. It has also marked out the achievement of the first Millennium Development Goal (MDG I) – halving hunger and poverty before 2015 – as the *sine qua non* for the achievement of the other ones.

As is well known, the great majority of the world's hungry and poor live in rural areas, mostly in countries that are largely agricultural. Many of these areas are trapped in an unsustainability spiral that is characterized by lack of access to external inputs, soil mining, decreasing land productivity, and chronic poverty. There is widespread consensus that achievement of MDG I first of all requires this spiral to be broken. Policy makers and economists are discussing policies and instruments to this purpose, but the way to go remains subject to extensive debate.

This situation has vital consequences for agricultural trade policies. Both in the Doha Round and in various bilateral trade negotiations, improving the situation in developing countries has been recognized as a primary aim. Nevertheless, there is increasing evidence that the least developed countries will benefit little, or even be hurt, by simple trade liberalization. Opinions on how to respond to this insight widely diverge, and a fruitful dialogue is often hampered by taboos and talking at cross-purposes. This situation led Professor Per Pinstrup-Andersen and Dr. Niek Koning to propose a round-table discussion to upgrade the scientific debate on the issue and enhance its policy relevance. Several views and analyses should be brought together to chart the common ground and the bases of divergence and to see how our understanding of the relation between agricultural trade policies and development could be improved. The idea of such a roundtable discussion, which would require an open and unbiased attitude from the participants, was widely accepted. Various scientific peers and trade policy experts were willing to contribute. The Dutch minister of Agriculture, Nature and Food Quality was willing to sponsor the initiative – a contribution that is gratefully acknowledged.

In this volume the various opinions, discussions, common conclusions and recommendations that have come out of this round table are presented. Thanks to the efforts of the participants (and some authors who were invited after the round table itself) and thanks to the open attitude of the organizers, Pinstrup-Andersen and Koning, it has become an unusual and stimulating book. Few other books present such a wide range of views on agricultural trade liberalization and the least developed countries. It is hoped that this constructive engagement between different views will contribute to more balanced discussions in future WTO and bilateral negotiations, and that it will inspire reforms that really help the poor and hungry in the world.

Rudy Rabbinge
Chairman, CGIAR Science Council
University Professor, Wageningen University, The Netherlands

CHAPTER 1

AGRICULTURAL TRADE LIBERALIZATION AND THE LEAST DEVELOPED COUNTRIES

Introduction

NIEK KONING[#] AND PER PINSTRUP-ANDERSEN[##]

[#] *Agricultural Economics and Rural Policy Group, Wageningen University, The Netherlands*
[##] *Division of Nutritional Sciences, Cornell University, Ithaca, NY, USA*

THE DEBATE

The agenda for the round of trade negotiations that started in 2001 was called the Doha Development Agenda. Its name was a signal to developing countries that their interests would be prioritized. Agriculture was at the core of the agenda. It is a vital sector in most developing countries, particularly the least developed ones, and it is heavily protected in high-income countries, resulting in severe trade distortions. The round started with high hopes that agricultural trade reform would strongly benefit the developing world. However, the negotiations were slow and experienced several breakdowns and an inability to meet deadlines. They ended in a state of deadlock in July 2006. However, we do not believe that the Doha Round is dead. While it may be stalled for a period of time, informal as well as more formal efforts are being made to continue the trade negotiations. The uncertain outcome makes the content of this book even more critical because it provides an input into the debate and decision-making as to whether the least developed countries should take policy action to protect themselves against the current trade-distorting policies in the OECD countries or prepare for a future with liberalized agricultural trade.

In view of the large potential gains that could be achieved by the multilateral reform of agricultural trade, the failure of such reform would be extremely unfortunate. While high-income countries stand to gain much from the removal of existing distortions in agricultural markets, developing countries would also benefit

N. Koning and P. Pinstrup-Andersen (eds.), Agricultural Trade Liberalization and the Least Developed Countries, 1–12.

greatly because of the tremendous importance of the agricultural sector in their economies. Removal of trade distortions would raise their earnings from agricultural exports. Besides, expanded opportunities for investment and trade would generate multiplier effects that would further enhance economic growth and poverty alleviation.

The multilateral reform envisaged by the Doha agenda was commonly equated with 'trade liberalization'. The precise meaning of this and how far it should go were subjects of debate. Economists have presented ambitious liberalization scenarios and estimated their benefits to the developing world to be around one percent of its GDP[1]. However, these *ex ante* estimates tended to disregard distributive effects and adjustment costs. They also ignored potential endogenous price fluctuations that might be strengthened by trade liberalization, and they made the strong assumption of a smooth transition of labour and capital between sectors (FAO 2006; Gérard et al. 2003; Polaski 2006)[2].

This book addresses a specific issue: how the least developed countries (LDCs) would fare under a trade liberalization scenario. Freer trade in agricultural commodities might be expected to benefit these countries because the agricultural sector plays such an important role in their economies. However, many of these countries are net importers of food and would therefore be negatively affected if trade liberalization were to result in an increase in their import prices. Moreover, trade liberalization would erode the value of the preferential access that many LDCs have to various OECD markets. Poor domestic infrastructure, limited access to credit and technology, and poor domestic agricultural policies in many of the LDCs would hamper their ability to benefit from expanded export opportunities. As a consequence, the benefits of international trade liberalization for developing countries could become concentrated in a few middle-income countries such as Brazil, Argentina and Thailand, while poorer countries stand to lose (see Polaski 2006 and the chapter by Yu in this volume). On the other hand, such negative expectations provoke the objection that they are based on static comparisons that ignore induced internal development effects. Expanded trade opportunities would induce investment in rural infrastructure, technology and institutions which, in turn, would promote agricultural development and economic growth in the LDCs. As a consequence, even if the immediate net effects for the LDCs were to be negative, the longer-term effects could still be positive.

The debate concerning the impact of trade liberalization on the LDCs coincides with the rising interest in why so many of these countries are trapped in stagnation. Bad governance, insufficient social capital and other socio-political conditions have often been identified as the major causes (e.g. Bates 1981; Collier and Gunning 1999; World Bank 1981). Many assumed that these conditions could be corrected through donor conditionality and structural adjustment. However, the results of these actions have proven to be rather disappointing. Renewed attention has been given to the complex poverty traps in which LDCs find themselves and which may produce institutional and political problems as endogenous results (e.g. Sachs 2005). Several observers doubt whether trade liberalization can remedy this situation (e.g. UNCTAD 2004; Östensson, this volume). Such doubts would be reinforced if the LDCs were to lose from trade liberalization, at least in the short run.

In order to provide an input into the debate about the impact of trade liberalization on the least developed countries, Cornell University, Wageningen University and the African Economic Research Consortium organized a workshop on "Agricultural Trade Liberalization and the Least-Developed Countries" at the end of 2004. Most of the chapters in this book are based on papers presented and discussed during that workshop.

Drawing on the chapters of the book, the rest of this introduction is organized around four key questions:

- Is agriculture an important driver of pro-poor economic growth in LDCs?
- Would LDCs gain from international agricultural trade liberalization?
- Should LDCs erect protective import tariffs?
- Why are there such large differences of opinion regarding the impact of agricultural trade liberalization on LDCs?

IS AGRICULTURE AN IMPORTANT DRIVER OF PRO-POOR GROWTH IN LEAST DEVELOPED COUNTRIES?

The common starting point of the authors of this volume is that most – if not all – LDCs need agricultural growth to get their economies moving. In this they follow Johnston and Mellor (1961), who reacted against the older idea that industrialization, based on taxation of agriculture, could do the job. In their seminal paper, these authors emphasize that agricultural growth provides opportunities for upstream and downstream activities as well as savings, labour and food (as a wage good) for other sectors. Follow-up research highlighted the effect on rural consumer demand for non-farm products (e.g. Block and Timmer 1994; Delgado et al. 1998; Hazell and Röell 1983). Empirical support came from Kuznet's (1966) demonstration that in developed countries, the onset of industrialization had almost always been linked with an economic revolution in the agricultural sector. Other studies confirmed the stimulating effect of agricultural growth on non-farm activity in developing countries (e.g. Liedholm et al. 1994) and the relative importance of domestic demand in early phases of their development (e.g. Balassa 1978; Heller and Porter 1978; Urata 1989). These and related findings were validated by experiences of several Asian countries, including China, Korea and Indonesia.

Nevertheless, after 1980 some economists began to doubt whether a domestic demand push was still a *conditio sine qua non* for development. In a globalized economy, they thought, export demand could be an alternative booster of growth. As a consequence, any suitable non-agricultural export sector could function as a promoter of development. In spite of the apparent logic of this reasoning, however, the success of the Asian tigers' industrialization was squarely based on rapid agricultural growth (e.g. Francks et al. 1999). This begs the question whether, in addition to the market-mediated linkages highlighted by Mellor and his followers, agricultural development might have vital non-market-mediated external effects on non-farm growth. Timmer (1995) suggests that effects on skills may be important. Social capital is another possible candidate (Koning 2002).

The authors in this volume agree that agricultural growth is indeed needed to start up non-farm development in many LDCs. However, achieving agricultural growth requires important conditions to be fulfilled. There is consensus that public investment in hard and soft infrastructure should increase. Without roads and suitable technologies, agricultural development will not take place. There is also agreement that effective market chains are needed to provide farmers with inputs and credit, assure quality and allow them to benefit from value added products. Rather than a simple dismantling of inefficient parastatal organizations, new forms of public–private cooperation may be needed to achieve this. Moreover, there is consensus that adequate farmgate price relations are needed to allow farmers to invest in sustainable agricultural intensification. This may require a further correction of domestic policies that exploit agriculture to pay the expenses of ineffective bureaucracies. Moreover, it requires the reform of agricultural trade policies in other countries as well as in LDCs themselves. There is agreement that OECD dumping is harming LDCs, and that what was emerging from the negotiations before they stalled would do little to improve the situation for LDCs. However, it is at this point that the consensus stops. The authors in this volume have widely divergent opinions on whether liberalization by OECD countries would be beneficial for LDCs, and on whether LDCs would be well-advised to protect their own farmers against cheap imports with which they cannot compete.

WOULD LEAST DEVELOPED COUNTRIES GAIN FROM INTERNATIONAL AGRICULTURAL TRADE LIBERALIZATION?

There is consensus that the ability of the LDCs to benefit from agricultural trade liberalization is limited by severe supply constraints. Nevertheless, some authors in this volume are convinced that, on balance, the LDCs would gain (see chapters by Badiane, Nassar, and Tutwiler and Straub). Other authors are less optimistic. Olle Östensson notes that LDC economies are already very open. Many small farmers do not produce export crops and are vulnerable to import competition. Poorly functioning markets, deficient rural infrastructure, and lack of access to technology and credit result in very low aggregate supply elasticities that prevent LDCs from taking advantage of expanded trade opportunities. Removal of OECD import tariffs is not a substitute for domestic investment for improving these conditions. Rather, the latter is a pre-condition for gaining from the former. Besides, private standards that are introduced by supermarkets may be more restrictive for LDC exports than public trade barriers.

Östensson expresses UNCTAD's view that the LDCs' dependence on commodities that receive low and unstable prices in world markets results in a poverty trap that involves generalized poverty, high population growth and under-funded governments, and which hampers diversification (see also UNCTAD 2004). In a similar vein, Kimsey Savadogo, and Andrew Dorward, Jonathan Kydd and Colin Poulton highlight more specific poverty traps that interact with trade policies. Savadogo focuses on the widespread soil degradation in Sub-Saharan Africa. In his view, extreme poverty makes parents discount the future of their children in their decisions on land management, causing an intergenerational tragedy of the commons even where land is bequeathed within families. Trade liberalization will not change this situation. Rather, massive development aid is needed to allow farmers to escape from this trap. Dorward and his colleagues focus on a low chain investment trap that hinders the development of supply and marketing chains that are needed for effective agricultural intensification. As a consequence, farmers and rural traders remain locked into 'atomistic relational market systems' rather than shifting to 'market and hierarchy reputational systems' as is needed for development (cf. Fafchamps 2004).

Daryll Ray and Harwood Schaffer question the validity of the widely held position that the decoupling of agricultural subsidies in OECD countries would reduce existing trade distortions. They believe that agricultural markets can only be balanced by supply management (and, on the multilateral level, managed trade). Because the shift to direct payments in the US was coupled to an abandoning of supply management, it did not reduce agricultural production but exacerbated the fall in international agricultural prices in the late 1990s. This prompted an increase in direct payments themselves and led to the current situation where this form of support has become an instrument of disguised dumping.

David Blandford and Wusheng Yu highlight the problem of preference erosion. LDCs have preferential access to various OECD markets where they benefit from relatively high prices. These benefits would decline if preference-providing countries were to reduce their own price supports. Yu presents results from analyses based on the latest version of the GTAP database. In his baseline liberalization scenario, where non-LDCs halve their MFN tariff rates of agri-food products, LDCs lose out because they receive lower prices for their exports. His results support those of Polaski (2006) and the arguments of Panagariya (2005), who also contend that preference erosion will cause poor countries to lose rather than gain by OECD liberalization.

Blandford and Yu believe that preferences should be increased to compensate LDCs for losses they incur as a result of international trade liberalization. Even if preferences turn out not to be sustainable in the long term, the continuing absence of a level playing field in international trade and the economic challenges facing the LDCs make an increase in preferences desirable in the short term. Additionally, preferences should be extended to a larger number of products, as well as being deepened by increasing the tariff concessions and quotas involved, or better still, by providing tariff-free and quota-free access as in the Everything-But-Arms initiative of the EU. Moreover, the number of preference-providing countries should increase and also expand to include middle-income countries. To illustrate this point, Yu compares his baseline liberalization scenario (where LDCs lose) with one where all advanced economies provide the Everything-But-Arms preferences of the EU. In the latter scenario, the negative outcome for LDCs disappears. He also presents a scenario where important middle-income countries provide the same far-reaching preferences. However, this appears to benefit LDCs less than open access to the markets of developed countries – a finding that contradicts the widespread belief that LDCs would especially benefit from more South-South trade.

Increasing preferences presupposes that the countries that provide them would not abandon their own price supports fully and immediately. This makes proponents of rapid and radical OECD liberalization wary of preferences. Ousmane Badiane contends that preferences are of little use to African countries because they are a pretext for maintaining OECD policies that are crippling the development of their export capacities. It may be questioned, however, whether OECD liberalization would indeed entail a quick release of supply constraints in LDCs, as he thinks. The experiences of the recent liberalization of the textile market point to another possibility: that a few well-placed countries would out-compete LDCs in their efforts to increase exports. Due to their advantage in scale and technology, countries like Australia and Brazil would rapidly be able to fill any new room that OECD liberalization were to create in the international market. André Nassar, who argues for liberalization from the viewpoint of one such country (Brazil), is critical of preferences. In his view, LDCs that are hurt by preference erosion should be compensated by the countries that provided the preferences – a viewpoint that deviates from the welfare economic principle that the winners of a policy reform (which include Brazil) should compensate the losers.

This controversy leads to another issue. Are all aspects of OECD protection equally harmful for agricultural development in the LDCs? Arguably, a distinction should be made between price support itself and the effect of this support on OECD production volumes. It is the latter that leads to import substitution and dumping, but in principle, this effect could be prevented by supply management. Thus, after the WTO sugar ruling, EU farm interests and ACP countries joined hands and demanded a further reduction in the production quotas for EU producers to avoid a strong reduction in prices. This issue is linked to that of the functioning of agricultural markets. The idea that liberalization would entail global gains (that could potentially benefit LDCs) presupposes that markets are able to balance demand and supply at prices that lead to equal marginal rates of transformation of farm products into other commodities. However, whether agricultural markets can also achieve this is subject to debate. A few decades ago, most agricultural economists agreed that this was not the case. They were convinced that the inelastic demand for food, and the way in which small-scale farm enterprises were encapsulated in the production-increasing environment of a modern industrial economy, were causing chronic oversupply and price instability (Schultz 1945; Cochrane 1958; Hathaway 1964; Tweeten 1970). The micro-economic revolution and the changed political climate after 1980 have altered this consensus (Gardner 1992), but there is still something unsatisfactory about the empirical foundation of this shift in opinion. Farm household income statistics that have been cited to disprove the existence of sub-normal farm earnings include off-farm incomes and are not corrected for sectoral differences in working hours (ibid.; also, e.g., Hill 2000; OECD 2003). Moreover, they refer to supported farm incomes whereas Schultz and his followers focused on farm income under free market conditions. Therefore, a minority of agricultural economists, including Ray and Schaffer and Koning in this volume, adhere to the older consensus. They believe that only a handful of countries with exceptional advantages in agriculture – like Australia, Brazil and Thailand – can hope to achieve normal earnings and a normal increase in productivity under global free trade. All other countries need some form of farm income support to achieve such aims. Balanced multilateral coordination of farm policies can only be based on a rationing of the market – managed trade, not free trade. 'Liberalization' is just an ideological smoke screen behind which the US and the EU are replacing one form of offensive protectionism with another. In this view, which also appears in the contribution by Sophia Murphy, the trade distortions caused by OECD farm policies should be redressed by restoring the original GATT principle that farm income support should be linked with supply management and export controls. Murphy concludes that LDCs have a strong interest in a multilateral, rule-based system of agricultural trade, but that the Doha agenda had little to offer to LDCs. Rather than concentrating on global deregulation, a real 'development round' should focus on stronger rules against dumping, stabilization of commodity prices, linking tariffs to supply management and export controls, stronger transparency requirements for large agribusiness companies, and the protection of social and environmental standards and of national development needs.

SHOULD LDCs ERECT PROTECTIVE TARIFFS?

Tariff protection by LDCs is generally accepted as an anti-dumping measure. This begs the question of what dumping exactly is: exporting below domestic prices, or below costs of production in exporting countries? The latter definition, which is advocated by Murphy, is relevant because the production costs of major export crops of the US and the EU are significantly above world market prices (Ray et al. 2003).

The real issue, however, is tariff protection that goes beyond anti-dumping. LDC tariffs on agricultural imports are generally low, and it might be asked whether raising them could help to get agriculture moving. Of course, the idea that protective tariffs might benefit LDCs is entirely at odds with equilibrium models that suggest that developing countries would benefit even more from reducing their own tariffs than from OECD liberalization. However, what if low tariffs lead to import competition of domestic producers while alternative earning possibilities are not forthcoming? This could occur if farmgate prices are a cog in poverty traps in which LDCs are caught. Accordingly, Dorward and his colleagues argue that price support and tariff protection might help to break the low chain investment trap in which farmers and rural traders are locked. In their view, tariff protection can be part of the policy package that is needed to shift LDC economies to a higher-level equilibrium. This would become even more important if OECD countries do not manage supply and allow their 'decoupled' support to lead to disguised dumping, as discussed by Ray and Schaffer and by Koning.

On the other hand, it should be remembered, as David Dawe remarks in his chapter, that agricultural tariffs are a regressive tax on consumers, including rural labourers and small farmers who are net buyers of food. Besides, higher food prices may raise wage costs affecting the competitiveness of the economies of LDCs. The question is whether these negative effects are outweighed by increases in farm employment and linkage effects on the non-farm economy. This is an empirical issue, which makes it useful to look at *ex post* experiences with trade liberalization or protection. Dawe looks at recent instances of agricultural trade liberalization in South and South-East Asia. Most occurred rather spontaneously, without being enforced by WTO decisions or other external forces. Apart from the reduction of tariffs on vegetable oil in India, they did not produce the tragic effects that NGO activists have depicted. In various cases, the effects were quite positive. Dawe also contrasts the experience of the Philippines, where increasing rice protection coincided with stagnating yields in the 1990s, with that of Thailand, where free trade coincided with continued yield growth. He refrains from drawing conclusions, though, as the stagnation in the Philippines could also have been due to an exhaustion of the Green Revolution that had not been fully introduced in Thailand at the time. Koning surveys historical experiences in OECD countries and Taiwan that could be relevant for LDCs. Between 1880 and 1930, agricultural free trade in Britain entailed total stagnation of productivity growth in agriculture, while in Germany the introduction of protection was followed by rapid farm progress that probably contributed to the country's rapid GDP growth. The German experience was repeated in Japan, South Korea and Taiwan in the Interbellum and the post-World War II period. Conversely, in Italy and France before World War I (WWI), agriculture stagnated in spite of protection.

These contrasting experiences suggest that other conditions determine whether the positive effects of agricultural import protection outweigh the negative ones. Perhaps innovation support, infrastructural policies and property rights can help to explain the difference. Germany before WWI led in the field of farm education and research, and underwent various phases of land reform. Similarly, the East-Asian countries benefited from far-reaching land reform, large investments in rural infrastructure and farm progress, and international green revolution research. Where such conditions are present, agricultural price support might accelerate farm progress, whereas without them, price support may achieve little more than a static redistribution of incomes. This latter may explain the slow increase in farm productivity in France and Italy before WWI, and possibly also that of the Philippine rice sector in the 1990s. The lessons to be drawn are that any protection that goes further than anti-dumping should be part of an encompassing policy package to increase productivity, including investments in rural infrastructure, research and technology, land reform and domestic market institutions.

It may further be noted that if LDCs were to introduce more systematic tariff protection, they could best do so in the frame of subregional customs unions. Internally, these could apply free trade to deter smuggling and allow specialization according to comparative advantage. Although most LDCs are currently net food importers, tariff protection could increase their food production beyond self-sufficiency. In that case, import tariffs alone would no longer be effective. This also means that raising tariffs in LDCs is no substitute for multilateral reform to improve agricultural prices in the world markets.

WHY DIFFERENCES OF OPINION?

The above discussions are influenced by different theoretical perspectives. In the standard neoclassical view, trade liberalization leads to global benefits. This view lies at the foundation of world trade models like GTAP. In fact, economists have a bad habit of calling the benefits indicated by such models 'welfare gains'. They are *potential* welfare gains (potential Pareto improvements), which only would become real gains if the winners were to compensate the losers (Koning and Jongeneel 1997).

Other theoretical perspectives question whether the standard model gives an adequate description of the real world. One of these perspectives is that of *new trade theory* (Krugman 1996; Krugman and Obstfeld 2006) and *endogenous growth theory* (Romer 1986; Lucas 1988). A central element of this is a virtuous cycle: investment in capital involves economies of scale or positive external effects that facilitate further growth. This perspective plays an important role in the 'hunt for large numbers': faced with the limited benefits of trade liberalization that are shown by static general equilibrium models, many economists are looking for economies of scale or endogenous growth mechanisms that would amplify the benefits of trade liberalization (e.g. Pack 1994; World Bank 1993). More simple dynamic amplifiers are built into some of the model studies that Ann Tutwiler and Matthew Straub cite to underline the benefits that agricultural trade liberalization would have for developing countries. A similar notion is also implicit in Badiane's idea that OECD liberalization will relax existing supply constraints in African countries. It should be noted, though, that new trade theory and endogenous growth theory also point to the possibility that any room for increased exports may be filled by well-placed countries like Australia and Brazil rather than by LDC producers.

A second perspective, and the logical counterpart of the previous one, is that of *poverty trap* (endogenous stagnation or decline). In this view, LDCs appear as multiple equilibrium systems that may be locked into low-level equilibrium. This perspective is central to the discussion by Savadogo, Dorward et al. and Östensson of problems in land management, chain investment and export specialization. The idea of multiple equilibriums leads to the necessity for a big policy push (which may or may not include tariff protection) to achieve a transition to a higher-level equilibrium.

A third perspective is that of dynamic sectoral disequilibrium. While the previous perspectives focus on local effects, this highlights the possibility that unbalancing forces inherent in economic growth dominate the equilibrating forces in global markets. This perspective lies at the heart of the idea of Ray and Schaffer and of Koning that agricultural markets are prone to chronic oversupply. Dynamic disequilibrium has some serious implications for standard trade models. The statistical data on market volumes and prices that are used for calibrating these models are seen as reflecting equilibrium. If they really pertain to a disequilibrium situation, the power of the model to predict the changes caused by an alteration of trade policies is compromised. Moreover, the 'benefits' shown by the models can no longer be interpreted as potential Pareto improvements.

Proponents of the above three perspectives welcome a more empirical approach to supplement the theoretical models that underlie *ex ante* assessments of the effects of trade liberalization. This would seem to resemble the old *Methodenstreit* between the Historical School and the (Neo-)Classical School in economics. However, the issue is not one of an inductive approach versus a deductive approach, but of a sound dialogue between theory and experience. This latter should go further than an improved empirical calibration of models; it should also include comparative case studies – the viewpoint chosen by Dawe and Koning in this volume.

SO WHAT IS THE BOTTOM LINE?

We, the editors, started out on this project because we were both convinced that investment in farm progress is vital to get LDC economies moving, and because we very much wanted an open discussion on the relation of trade policies with this standpoint, on which we held different opinions. Having participated in the workshop discussions and having read the papers presented in this volume, as well as other relevant literature, we agree on the following conclusions:

1. Agricultural and rural development continues to be the strongest driver of broad-based economic growth and poverty alleviation in LDCs. However, for this driver to generate the multiplier effects essential for achieving pro-poor growth, LDCs must increase investment in rural infrastructure, research and technology, as well as designing and implementing policies that will facilitate appropriate institutions and well-functioning input and output markets. Without such investments and policy changes, any benefits gained from the OECD liberalization of agricultural trade will go primarily to middle and high income countries, leaving LDCs with little or no benefits. In the short run, LDCs may in fact lose out.
2. Without some form of extra-market supply management, removal of trade-distorting OECD policies is unlikely to reduce OECD supplies as much as predicted by current models. Thus, price increases in the international markets may be less than expected, as will export opportunities for developing countries.
3. LDC tariff reduction is a very poor substitute for multilateral reform to remedy trade-distorting policies by OECD countries.
4. Anti-dumping tariffs may be appropriate for LDCs to protect domestic agriculture against imports with prices that are significantly below the costs of production in the exporting countries. Such tariffs would be limited to the size and duration in which the (thus defined) dumping would occur.
5. If LDCs were to choose to apply protective import tariffs that go beyond anti-dumping, they should couple this to very significant increases in investments in domestic markets, infrastructure, research, technology and other public goods, and with policy changes in favour of agricultural and rural development. Without such policies, protective import tariffs could strongly harm low-income consumers, and risk stifling economic growth and productivity increases in the non-agricultural sector.

6. In the absence of multilateral reforms, preferential arrangements for LDCs should be increased, and safeguards and exclusion of sensitive products from special agreements such as the EBA should be removed. Truly free access to OECD markets for all commodities and products from LDCs in unlimited quantities should be granted. This would include the elimination of tariff escalation for all processed agricultural commodities.

NOTES

[1] Somewhat larger gains that World Bank economists presented at one point had to be revised (World Bank 2003; Anderson and Martin 2005).

[2] Only a part of these objections is answered by more elaborate model studies that try, for example, to include the differential effects on different categories of households (Hertel and Winters 2005).

REFERENCES

Anderson, K. and Martin, W., 2005. *Agricultural trade reform and the Doha development agenda.* World Bank, Washington. World Bank Policy Research Working Paper no. 3607.

Balassa, B., 1978. Exports and economic growth: further evidence. *Journal of Development Economics,* 5 (2), 181-189.

Bates, R., 1981. *Markets and states in tropical Africa: the political basis of agricultural policies.* University of California Press, Berkeley.

Block, S. and Timmer, C.P., 1994. *Agriculture and economic growth: conceptual issues and the Kenyan experience.* Harvard University, Boston. HIID Development Discussion Paper no. 498. [http://www.cid.harvard.edu/hiid/498.pdf]

Cochrane, W., 1958. *Farm prices: myth and reality.* University of Minnesota Press, Minneapolis.

Collier, P. and Gunning, J.W., 1999. Explaining African economic performance. *Journal of Economic Literature,* 37 (1), 64-111.

Delgado, C.L., Hopkins, J. and Kelly, V.A., 1998. *Agricultural growth linkages in Sub-Saharan Africa.* IFPRI, Washington. IFPRI Research Report no. 107. [http://www.ifpri.org/pubs/abstract/107/rr107.pdf]

Dorward, A., Kydd, J., Morrison, J.A., et al., 2002. *A policy agenda for pro-poor agricultural growth: paper presented at the Agricultural Economics Society Conference, Aberystwyth, 8th-10th April 2001.*

Fafchamps, M., 2004. *Market institutions in Sub-Saharan Africa: theory and evidence.* MIT Press, Cambridge.

FAO, 2006. *Towards appropriate agricultural trade policy for low income developing countries.* FAO, Rome. FAO Trade Policy Technical Notes on Issues Related to the WTO Negotiations on Agriculture no. 14. [ftp://ftp.fao.org/docrep/fao/009/j7724e/j7724e00.pdf]

Francks, P., Boestel, J. and Kim, C.H., 1999. *Agriculture and economic development in East Asia: from growth to protectionism in Japan, Korea and Taiwan.* Routledge, London.

Gardner, B.L., 1992. Changing economic perspectives on the farm problem. *Journal of Economic Literature,* 30 (1), 62-101.

Gérard, F., Piketty, M.G. and Boussard, J.M., 2003. Libéralisation des échanges et bien-être des populations pauvres: illustration à partire du modèle ID^3 de la faiblesse des impacts et de la sensibilité des resultats aux hypothèses de fonctionnement des marchés. *Notes et Études Économiques no. 19,* 111-134.

Hathaway, D.E., 1964. *Government and agriculture: public choice in democratic society,* New York.

Hazell, P.B.R. and Röell, A., 1983. *Rural growth linkages: household expenditure patterns in Malaysia and Nigeria.* International Food Policy Research Institute, Washington. IFPRI Research Report no. 41.

Heller, P.S. and Porter, R.C., 1978. Exports and growth: an empirical re-investigation. *Journal of Development Economics,* 5 (2), 191-193.

Hertel, T.W. and Winters, L.A., 2005. *Putting development back into the Doha agenda: poverty impacts of a WTO agreement.* World Bank, Washington.

Hill, B., 2000. Agricultural incomes and the CAP. *Economic Affairs,* 20 (2), 11-17.

Johnston, B.F. and Mellor, J.W., 1961. The role of agriculture in economic development. *The American Economic Review,* 51 (4), 566-593.

Koning, N., 2002. *Should Africa protect its farmers to revitalise its economy?* International Institute for Environment and Development, London. IIED Gatekeeper Series no. 105.

Koning, N. and Jongeneel, R., 1997. *Neo-Paretian welfare economics: misconceptions and abuses.* Wageningen University, Wageningen. Wageningen Economic Papers no. 05-97.

Krugman, P.R., 1996. *Pop internationalism.* MIT Press, Cambridge.

Krugman, P.R. and Obstfeld, M., 2006. *International economics: theory and policy.* 7th edn. Pearson/Addison-Wesley, Boston.

Kuznets, S., 1966. *Modern economic growth: rate, structure, and spread.* Yale University Press, New Haven.

Liedholm, C., McPherson, M. and Chuta, E., 1994. Small enterprise employment growth in rural Africa. *American Journal of Agricultural Economics,* 76 (5), 1177-1182.

Lucas, R.E., 1988. On the mechanics of economic development. *Journal of Monetary Economics,* 22 (1), 3-42.

OECD, 2003. *Farm household income: issues and policy responses.* OECD, Paris.

Pack, H., 1994. Endogenous growth theory: intellectual appeal and empirical shortcomings. *The Journal of Economic Perspectives,* 8 (1), 55-72.

Panagariya, A., 2005. Agricultural liberalisation and the least developed countries: six fallacies. *The World Economy,* 29 (8), 1277-1299.

Polaski, S., 2006. *Winners and losers: impact of the Doha Round on developing countries.* Carnegie Endowment for International Peace, Washington. [http://www.carnegieendowment.org/files/Winners.Losers.final2.pdf]

Ray, D.E., De la Torre Ugarte, D.G. and Tiller, K.J., 2003. *Rethinking US agricultural policy: changing course to secure farmer livelihoods worldwide.* Agricultural Policy Analysis Centre, University of Tennessee, Knoxville. [http://apacweb.ag.utk.edu/blueprint/APAC%20Report%208-20-03%20WITH%20COVER.pdf]

Romer, P.M., 1986. Increasing returns and long-run growth. *The Journal of Political Economy,* 94 (5), 1002-1037.

Sachs, J., 2005. *The end of poverty: how we can make it happen in our lifetime.* Penguin, London.

Schultz, T.W., 1945. *Agriculture in an unstable society.* McGraw-Hill, New York.

Timmer, C.P., 1995. Getting agriculture moving: do markets provide the right signals? *Food Policy,* 20 (5), 455-472.

Tweeten, L., 1970. *Foundations of farm policy,* Lincoln.

UNCTAD, 2004. *The least developed countries report 2004.* UNCTAD, New York. [http://www.unctad.org/en/docs/ldc2004_en.pdf]

Urata, S., 1989. Sources of economic growth and structural change: an international comparison. *In:* Williamson, J.G. and Panchamukhi, V.R. eds. *The balance between industry and agriculture in economic development. 2, Sector proportions.* MacMillan Press, London, 144-166.

World Bank, 1981. *Accelerated development in Sub-Saharan Africa.* World Bank, Washington.

World Bank, 1993. *The East Asian miracle: economic growth and public policy.* Oxford University Press, Oxford.

World Bank, 2003. *Global Economic Prospects 2004: realizing the development promise of the Doha Agenda.* World Bank, Washington.

CHAPTER 2

AGRICULTURAL TRADE, DEVELOPMENT PROBLEMS AND POVERTY IN THE LEAST DEVELOPED COUNTRIES

An overview

OLLE ÖSTENSSON

Head of Diversification and Natural Resources Section, Commodities Branch, UNCTAD, Switzerland

INTRODUCTION

The present paper attempts to give a broad overview of the situation of LDCs concerning the role of agricultural exports in economic growth and poverty reduction[1]. 'Trade policies' in this context are taken to include not only approaches to trade negotiations in the WTO or other fora, but also policies that target competitiveness and the capacity to export.

LDCs AND COMMODITY DEPENDENCE

The majority of LDCs (31 of the 49 countries so classified) depends mainly on commodities, particularly agricultural products (20 of the 31), for export earnings, and this dependence shows little sign of diminishing. Agriculture is the dominant economic activity in terms of employment in almost all LDCs, with on average 69% of the labour force engaged in agriculture in 2002, as compared to 54% in all developing countries (UNCTAD 2004b, Annex table 3).

Real commodity prices quoted for international markets generally exhibit a long-term downward trend in real terms. As can be seen from Table 1, this is true for all agricultural products over the last four decades. The tropical beverages group, which includes coffee, cocoa and tea, fares the worst. These products are also grown by small and poor farmers in many developing countries. The vegetable oilseeds and oils group, which faced a critical price situation comparable to tropical beverages 10

N. Koning and P. Pinstrup-Andersen (eds.), Agricultural Trade Liberalization and the Least Developed Countries, 13–24.

years ago, has recovered some of its losses in prices.

***Table 1**. Commodity prices 1964-2004*

Product group	Annual indices of monthly averages (1985=100)							
	Current $			Current SDRs		Real prices*		
	1964	1994	2004**	1994	2004**	1964	1994	2004**
Tropical beverages	33	91	54	63	37	89	58	36
Other food	66	152	131	106	90	178	96	87
Vegetable oilseeds and oils	46	107	111	75	77	124	67	74
Agricultural raw materials	46	140	126	98	87	124	89	83
Minerals, ores and metals	*49*	*124*	*151*	*86*	*104*	*132*	*79*	*100*

Source: UNCTAD, *Commodity Price Bulletin* various issues
* Current prices in dollars deflated by Manufactures Export Unit Value Index for developed countries
** August 2003-September 2004 average

As a result of the decline in real prices, commodity-dependent LDCs have generally experienced falling terms of trade. World Bank estimates for non-oil-exporting countries in sub-Saharan Africa, most of which are LDCs, suggest that their cumulative terms of trade losses over the period from 1970 to 1997 amounted to 119% of regional GDP in 1997 (Dehn 2000).

In general, commodity prices are inherently more volatile than prices of manufactured products[2]. Apart from lags in (the often imperfect) supply response to price signals and the impact of weather conditions, speculative activity also generates price fluctuations. Price fluctuations have continued to be a characteristic common to almost all commodity markets, and if anything, their amplitude appears to have increased (UNCTAD 2003, p. 3). This can be seen in Table 2, which gives percentage variations of the average monthly price around the exponential trend for selected price indices over two three-year periods.

The effects of commodity price instability are particularly significant in the LDCs since the scale of price shocks in relation to domestic resources available to finance investment, or savings, is extremely large. In a sample of 18 non-fuel-commodity-exporting LDCs for which data were available, the maximum two-year terms of trade shock over the period 1970-1999 led to income losses of over 100% of the domestic resources available to finance investment in any given year in eight of them, and income losses of over 25% of domestic resources available to finance investment in a further eight (UNCTAD 2000, pp. 38-39). Negative price shocks have a negative effect on economic growth, particularly through their impact on the utilization of productive capacity, and there is not a similar offsetting positive effect

Table 2. *Instability indices for prices: commodity groups and selected products*

	1989-92	2000-03
Tropical beverages	8.2	9.1
Coffee	*12.9*	*11.2*
Cocoa	*9.1*	*14.1*
Other food	3.6	4.0
Wheat	*12.1*	*7.0*
Sugar	*12.1*	*15.0*
Bananas	*17.7*	*20.3*
Vegetable oilseeds and oils	6.7	9.3
Soya bean oil	*3.5*	*9.1*
Palm oil	*10.2*	*11.3*
Agricultural raw materials	3.6	6.0
Cotton	*10.5*	*15.3*
Rubber	*5.7*	*13.2*

Source: UNCTAD secretariat calculations

from positive commodity price shocks. Among the macroeconomic mechanisms that have been found to be important causes of reduced growth due to negative price shocks are increasing real exchange rate instability, which leads in particular to poor resource allocation and lower factor productivity, and increasing fiscal instability, which contributes to the build-up of indebtedness and reduces the level of, and return on, investment.

The growth rate of LDC commodity exports has generally been disappointing and the share of these countries in world non-fuel-commodity exports has fallen dramatically over the last few decades, from 5.6% in 1966-1970 to 1.1% in 2000 (UNCTAD 2004a). Part of the explanation for this development is that the relative importance of different product groups in international agricultural trade is changing, and suppliers that have been able to position themselves in the trade of dynamic items are doing much better than others, including, in particular, LDCs, who are trapped in traditional items with stagnant trade and low value-added. For example, cereals, which had a share of around 12% in world agricultural trade thirty years ago, now have a share of barely 7%. Looking at developing countries as a whole, coffee, sugar and cotton, which were the top agricultural exports thirty years ago, have now been replaced by fish and vegetable oils. The bulk of the increase in dynamic exports has originated from the more advanced and already diversified countries of South-East Asia and Latin America. These countries have not only entered markets of non-traditional products, but have also added value to their exports, for example by supplying ready-made flower bouquets and vegetables that have been packaged and bar-coded and are ready to be put on the retailers' shelves. Even among LDCs, there are significant differences between commodity economies and manufactures and/or service exporters. The share of dynamic agricultural products (those with an annual percentage growth of world imports above the average nominal growth rate of total world imports from 1994 to 1998) in agricultural exports of the latter group of countries increased from 37% in 1981-83 to 48% in 1997-99, while for non-oil commodity exporters it only rose from 13 to 14% (UNCTAD 2002b, p. 147).

Another important change is the significantly faster increase in world exports of processed agricultural products than those of semi-processed and unprocessed agricultural products. Between 1990-91 and 2001-02, the share of processed products rose from 42% to 48% of global agricultural trade (WTO 2004, p. 17). However, the share of processed commodities in total LDC exports fell from 21 to 8% between 1981-83 and 1997-99 (UNCTAD 2002b, p. 147). Thus, in terms of domestic processing, instead of moving up the value chain, the LDCs are sliding down it.

COMMODITY DEPENDENCE AND POVERTY

The combination of falling real prices and slow export volume growth has led to stagnant or falling incomes in the agricultural sector of LDCs dependent on agricultural exports, and to foreign exchange shortages. Import volumes are low, and low levels of technology imports and lack of complementary imports result in a reduced level of investment, reduced efficiency of resource use and outdated production processes. With little surplus available for investment, either in the sector itself or in the public sector (which is responsible for providing necessary services such as infrastructure), productivity growth has remained low and these LDCs have fallen into a poverty trap. As described in UNCTAD (2002b, pp. 148 and 150) five main interrelationships constitute the domestic aspects of the poverty trap. All of them inhibit diversification into more dynamic products. First, domestic resources available to finance physical and human capital investment and productivity growth are low owing to generalized poverty. Second, state capacities are weak as all activities, including administration and law and order, are underfunded. Third, corporate capacities are weak, even though there may be a thriving informal sector. Fourth, generalized poverty engenders rapid population growth and environmental degradation. Fifth, in a situation of generalized poverty, the probability of political instability and conflict is greater.

The importance of commodity dependence as a determining factor for poverty is demonstrated by the difference in the incidence of extreme poverty (percentage of population living on less than 1$/day) between the different categories of LDCs. In manufactures exporters, this incidence was 25% in 1997-99 and in service exporters it was 43%. In agricultural exporters, it was 63% (and in mineral exporters it was 82%) (UNCTAD 2004b, chart 19, p. 132). In agricultural LDCs, which generally are not very urbanized, extreme poverty is mainly a rural phenomenon that affects those engaged in agriculture.

LDCs AND TRADE LIBERALIZATION

As is clear from the preceding argument, export dependence on a few commodities with slowly growing markets and declining real price trends has been a major constraint on diversification and growth in agricultural LDCs. Trade liberalization has done little to alleviate the problem. It should first be noted that the LDC economies are generally open. In 2002, of 46 LDCS for which data were available,

- the average tariff rate of 42 was less than 25%,
- the average tariff rate of 36 was less than 20%,
- the average tariff rate of 23 was less than 15%,
- in 29 LDCs, non-tariff barriers (NTBs) were absent or insignificant (less than 1% of production and trade was subject to NTBs), and
- in 28 LDCs, there were no or insignificant NTBs, and average tariff rates were below 25% (UNCTAD 2004b, p. 179).

Trade as a share of GDP in LDCs was 50.7% in 1999-2001, only slightly lower than in other developing countries and higher than in high-income OECD countries (UNCTAD 2004b, chart 13, p. 107). The IMF trade-restrictiveness index[3] is lower for LDCs than for other developing countries (UNCTAD 2004b, chart 32, p. 180). Trade liberalization is deeper in African LDCs than in Asian ones and also in commodity-exporting LDCs than in manufactures and service exporters (UNCTAD 2004b, p. 180-181).

The deep trade liberalization undertaken by agricultural exporters among the LDCs could have been expected to lead to higher growth, particularly taking into account the market access preferences that are extended to LDCs by other countries. However, as already noted, export growth has been modest and LDCs have lost market shares for agricultural commodities. Developed country agricultural protectionism and export support have obviously played a large role in hindering export growth in agricultural LDCs. Subsidized exports from developed countries affect both domestic and international markets and exert a negative influence on the diversification of production and exports from commodity-dependent countries. Loss of competitiveness relative to subsidized agriculture discourages investments in agriculture and local processing in non-subsidizing countries. Although urban consumers may enjoy access to cheaper food products, there are abundant examples of developing countries' products being displaced on domestic markets by imported ones from developed countries providing generous subsidies, and export markets being lost to suppliers from the same countries.

Although developing countries have been accorded preferences under a multitude of agreements[4], exceptions to these preferences often relate to agricultural products. For example, the European Union's initiative on 'Everything but Arms' (EBA) offers free market access to LDC products, with less than 5% of pre-EBA exports left facing a tariff barrier. According to simulations however, the impact of this initiative will be a relatively small increase in exports from the LDCs, as 70% of the potential positive trade effects would have come from free access for sugar, rice and beef, which has been deferred until 2006 (UNCTAD and Commonwealth Secretariat 2001). Moreover, utilization ratios for the preferences have generally been low. In 2001, only 68.5% of total imports from LDCs eligible to enter Quad markets at a preferential duty rate actually did so. The rest paid MFN duties. The low utilization ratios are mainly the result of the insignificant magnitude of potential commercial benefits; the lack of technical knowledge, human resources and institutional capacity to take advantage of preferential arrangements; and the conditions attached to the preferences. The effective benefits are significantly limited by their unpredictability and by rules of origin and product standards (UNCTAD 2004b, p. 250). Finally, it is also important to note that the conditions of access to WTO for recent entrants, such as Cambodia, Nepal and Vanuatu, have, if anything, been less favourable than the special and differential treatment accorded to developing countries, including LDCs, that are already WTO members (UNCTAD 2004b, p. 60-61).

TRADE LIBERALIZATION AND POVERTY

While positive effects in terms of export growth could of course be expected from multilateral trade liberalization, more interesting from the point of view of the present paper is their impact on poverty. According to calculations carried out by the UNCTAD secretariat, multilateral trade liberalization would slow down the rate of increase in the number of extremely poor people in the LDCs. Instead of increasing from 334 million in 2000 to 471 million in 2015 in the case of no liberalization, the number of poor would increase to 'only' 463 million (UNCTAD 2004b, p. 222). Part of the reason for this somewhat disappointing result is that most of the poor in the agricultural LDCs live in rural areas and are engaged in subsistence farming of traditional food crops. Improved export conditions can have an impact on the living conditions of this group if they shift their production mix. But such a production shift is not always possible, due to risk aversion and uncertainty. This group also will not benefit much from a reduction in import prices of consumer goods as the import content of their expenditures is very low. Moreover, if liberalization leads to a substitution of traditional food by imported products, the traditional producers may face declining demand and prices for their produce.

EFFECTS OF DOMESTIC LIBERALIZATION ON COMMODITY PRODUCTION AND EXPORTS

It appears that the direct benefits of multilateral trade liberalization would not be sufficient to make a significant impact on poverty in agricultural LDCs. The expected growth in exports would not be large enough to make a real dent in poverty. Moreover, the positive relationship between export growth and output growth appears to be weaker in LDCs than in other developing countries and export growth is not necessarily inclusive. Neither does domestic liberalization in agricultural LDCs, usually undertaken as part of structural adjustment programs, appear to have had unambiguously positive effects on commodity production and exports. Hopes that the action of market forces would lead to greater efficiency have not been fulfilled, partly because of the absence of functioning markets, partly because of unrealistic expectations of what markets can achieve under the best of circumstances.

The impact on the farming sector of one significant element of structural adjustment programs, namely abolishing marketing boards[5] and other governmental support structures for agriculture, including the liberalization of agricultural credit systems, has been generally negative. Although in many cases marketing boards were the instruments of an implicit taxation of the farming sector and suffered from inefficiencies and sometimes corruption, they also provided useful services. In many countries, the private sector has been unable to fill the gap and supply these services satisfactorily, basically as a result of its underdevelopment, and of unfavourable institutional, legal and regulatory frameworks. These services include the provision of information, finance and inputs as well as quality control. For example, in the cocoa market, cocoa from Ghana, which, unlike many other developing countries, has retained its marketing board, enjoys a quality premium, because of the market's confidence in quality assurance by the government.

Managing exposure to world-market price risks and holding products in storage to avoid losses and benefit from seasonal price variations are among the many new challenges for small farmers in dealing with a liberalized market where government support has been discontinued. The demise of governmental finance has also exacerbated the lack of working capital and poor access to credit for small-scale producers (in part as a result of smallholders not having viable collateral and the widespread inability of local banks to secure agricultural loans against, for example, future sales or commodity inventories).

In the first years of liberalization of domestic agricultural markets, some of the activities of the former government marketing boards were taken over by a range of local traders. Relatively quickly, however, international trading companies or their agents, and foreign traders with easy access to finance replaced these traders. Foreign firms, in particular large ones, were able to reach deep into the production, trading and processing levels in these countries. Anecdotal evidence on the impact of these changes on small farmers can be contradictory (UNCTAD 1999a). In some cases farmers were paid promptly and in cash, and enjoyed a slight increase in their share of (in most cases, a declining) world price. On the negative side, however, input use declined and the quality of the product fell. Nevertheless, especially in cases where intermediaries without an established presence in commodity markets act between small producers and large traders, the market does not seem to function and prices received by the farmer fluctuate almost randomly, thus losing their economic meaning.

One would expect that without the protective mechanism against price instability at the producer level that existed with marketing boards, producers would be directly subject to the fluctuations in international markets. It is not uncommon, however, for producers to face price instability much larger than that in the international price and even totally unrelated to it. In October 2004, a small cocoa farmer in Cameroon explained to an UNCTAD team that for one kilo of cocoa, he was paid 100 CFA in 2002, 800 CFA in 2003 and 400 CFA in 2004. The average international prices during the cocoa purchasing period in Cameroon in the corresponding years were 1224 CFA, 1182 CFA and 1254 CFA, respectively. Not only is the variation in the local price much larger than that in the international price, even the direction of change is contradictory. Better functioning and transparent markets could reduce the haphazard movements in domestic prices. Price risk management instruments are powerful tools for coping with price instability. But any attempt to use them is fraught with special difficulties under these circumstances. A well functioning domestic market with transparent links with international markets is a necessity in this respect.

SUPPLY-SIDE OBSTACLES AND MARKET ENTRY PROBLEMS

The strategic problem facing agricultural LDCs is how to exploit the market access accorded, whether or not on preferential terms, and convert it into export growth, diversification and, eventually, broad-based development and poverty reduction. LDC producers have lacked the competitiveness and supply capacities necessary to exploit both their comparative advantages and the potential advantage accorded by preferences. Thus, while market access has been assured, at least to some extent, market entry has not been so.

There are both supply-side and demand-side reasons for the lack of export success. Common supply-side obstacles include deficiencies in infrastructure and extension services, and lack of access to credit, technology and market information. Low productivity is rampant in the agriculture of many developing countries, especially in Africa. For example, maize yields are 1.6 tonnes per hectare in Africa compared to 3.8 tonnes per hectare in Asia (Sachs and Sanchez 2004), and the gap is not narrowing. Over the period 1980-1997, crop yields for seven agricultural exports were on average lower in LDCs than in other developing countries in all cases but cocoa (UNCTAD 1999b, table 23).

On the demand side, requirements of importers have become increasingly stringent, partly because of consumer preferences (health concerns, traceability) and partly as a result of restructuring, which has led to an increasing dominance of importing markets by large distribution networks. LDC producers, particularly the smaller ones, find it very difficult to meet these requirements, since they lack funds to undertake the necessary investments. In addition, importers' transaction costs when dealing with many small producers are high.

One important shift in multilateral trade negotiations that started with the Uruguay Round is the advent of positive rule making. Areas that, although linked to trade, have traditionally been the domain of domestic policies, are now part of international trade rules. Sanitary and phytosanitary measures (e.g. traceability), intellectual property rights (e.g. what seeds can be used – exports of roses from India to France were returned because of uncertified plant use), and operation of state trading organizations all have to conform to the outcome of the 'liberalization' process.

The General Agreement on Trade in Services of WTO aims to liberalize the retail sector. Although it cannot be said to be a result of this agreement, as service sectors have been opened up in line with liberalization, global supermarket chains have increased their dominance of the retail sector in many countries. The requirements for supplying supermarkets are different from those of selling in traditional markets. Here again, while the playing field is becoming more level, rules are becoming more complicated, and only those that can play according to these rules can take part in the game. Small producers everywhere are finding it more difficult to meet the requirements but those in LDCs are further disadvantaged not only because they are poorer, but also because the requisite institutional structure and governmental support are lacking.

The increasing presence of international distribution firms and supermarket chains in food trade and the retail sector has generated significant impacts on small-scale farmers not only in developing countries but also in developed ones. Their growth and dominance are reflected in the marked surge of foreign direct investment flows into the retailing sector (Reardon and Berdegué 2002, p. 376). With the advent of the 'global supermarket', the distinction between world and local markets is fast disappearing. Quality concerns and modern business practices reminiscent of the international markets are being transferred and diffused into domestic markets. This becomes even more so as markets are opened up and competition with imports becomes inevitable. This is true not only in developed and relatively richer developing countries, but also in Africa (Weatherspoon et al. 2003). Those who can meet the requirements of supermarkets in international markets, and only those, are likely to succeed in the higher segments of domestic markets. Efficiency gains imposed by meeting the standards may also lead to higher earnings for the successful farmer.

Small and LDCs producers are at a disadvantage under these new trading practices since not only *what* is produced, but also *how* and *by whom* it is produced emerge as important concerns. Firstly, the simple understanding of the exigencies is a complicated matter. Secondly, meeting these exigencies requires investments that small producers are usually unable to undertake individually. Investments for meeting health, safety and quality requirements can range from upgrading management skills to purchasing new equipment and establishment of quality control and coordination systems. Therefore the importance of cooperative action among small producers is evident. The large size of the importers, coupled with the necessity to ensure quality, traceability and continuity in supplies bestows an advantage for large farms over smaller ones, stemming from lower transaction costs. This is another reason for small farmers to organize themselves and act cooperatively.

While market access barriers and international trade measures implemented by governments comprise the first hurdle to selling in international markets, clearing this hurdle does not guarantee that a product will appear on retailers' shelves. For instance, SPS requirements define the conditions necessary but not those sufficient for being able to export. Many, and in most cases much more stringent, quality and labelling requirements, as well as conditions regarding production and processing practices are imposed by importing firms either individually or collectively as is the case with Eurepgap (http://www.eurep.org/). Particularly in the case of food items, meeting the requirements of importing firms and distribution and retailing channels is the principal prerequisite for success, and the burden for doing so ultimately falls upon the farmer. These requirements are usually more stringent than the government regulations reflected in measures undertaken in accordance with the requirements of the SPS Agreement. Moreover, when requirements are imposed by private enterprises, there is no way to contest them legally, except in situations where rules on competition are violated (UNCTAD 2002a, p. 11).

Accordingly, while improvements in market access and a dramatic reduction in agricultural protectionism are necessary conditions for improving the export performance of LDCs, they are not sufficient conditions. More needs to be done to raise the competitiveness of LDC producers, and this will require massive investment in the upgrading of capacities in both the public and the private sector.

A 'WINDOW OF OPPORTUNITY'

One of the more promising developments in recent years is the emergence of a new dynamic element in international commodity trade: rapidly increasing Asian demand. Together with improved export opportunities that could result from changes in the international trading system, increased demand for commodities in these countries could considerably boost world demand for commodities. A 'window of opportunity' could thus open up over the next several years, allowing substantially improved export earnings for developing countries.

Commodities have accounted for a constant (in the case of China) or increasing (in the case of India and the rest of Asia) portion of total imports, which are rising rapidly. Moreover, the share of Asian commodity imports coming from other developing countries has increased steadily. In China, agricultural imports from developing countries outside Asia increased by a total of 30% from 1995 to 2002. Since this was a period when commodity prices decreased dramatically, the growth in volume terms was higher. China's share of world consumption of food products is increasing fast. Its major agricultural commodity imports are cereals (mostly wheat and barley); vegetable oils and oilseeds (particularly soybean, soybean oil and palm oil); fish and seafood, and animal feed. Other growth sectors include horticultural products such as cut flowers and fruits and vegetables, as would be expected at the present level of income.

What are the implications of continued rapid growth in China and India for world commodity demand? In order to answer this question, two important facts need to be kept in mind. First, China and India have a combined population of 2.3 billion people, about 37% of the world's population. Thus, a US$100 increase in the per-capita income of these two countries (10% for China and 20% for India) represents US$ 230 billion in additional demand. Second, both countries are at a stage of industrialization (with China somewhat ahead of India) where per capita commodity consumption tends to increase rapidly and they are likely to remain at this stage for the next few years. Lifestyle changes, including in dietary patterns, brought on by rising income and urbanization, will change the composition of demand for food products. As just mentioned, this is already reflected in the composition of China's agricultural imports, and also in India's, although to a smaller extent. Since both countries have made great progress in reducing poverty, this last factor is of major importance since, other things being equal, rising incomes for the poorer segments of the population tend to have a large impact on food consumption. Imports of non-traditional products such as coffee and cocoa are also increasing, and this development is important for other developing countries.

Some of the increased demand for commodities will be met from domestic production. With respect to agricultural goods, productivity improvements in both countries would be expected to lead to increasing agricultural production. However, in both China and India, shortages of arable land may prove to be a constraint on production increases. The amount of land devoted to rice production in China has declined over the last six years, and the rice harvest is expected to fall to a twelve-year low of 126 million tonnes in the current crop year. Within a decade, China may have to import up to 50 million tonnes of grain per year.

Accordingly, China and India can be expected to need increasing commodity imports for several years to come. For reasons of geographical proximity, Asian countries could be expected to be major beneficiaries, and Chinese demand has already led to improved markets for its neighbours. However, other developing regions are also beginning to see major increases in their commodity exports to China. African countries experienced a 10% annual increase in agricultural exports to China from 1995 to 2002, and Latin American countries saw exports increase at an annual rate of over 4%, albeit in both cases from relatively low initial levels.

In conclusion, Chinese and Indian demand growth will provide a major dynamic stimulus to international commodity markets over the next few years, and since commodity markets are global, the additional demand arising from Asian growth will benefit a wide range of countries and not only affect their immediate neighbourhood. Although problems of oversupply of individual commodities will continue to affect export earnings and producers' incomes, the general trend should be positive.

CONCLUSIONS

LDCs depending on agricultural exports have experienced low growth and shrinking market shares for the last few decades. With little surplus available for investment, productivity growth has remained low and these LDCs have fallen into a poverty trap. Domestic liberalization has had little significant positive effect on export growth and poverty. Multilateral trade liberalization would only yield very limited results in terms of poverty reduction. Sufficient growth in exports to have a tangible effect on poverty would require the removal of supply-side obstacles as well as measures to facilitate the meeting of market exigencies. A 'window of opportunity' for LDCs to increase their earnings from agricultural commodity exports may be opening as a result of increases in commodity demand in Asia, particularly China and India. This assumes, however, that developed countries do not use the expected demand increase as an excuse for taking a complacent view of the need for reductions in agricultural tariffs or support to domestic producers; or worse, that they exploit the market growth for their own exclusive benefit by continuing export subsidies and high levels of domestic support. On the other hand, a prolonged period of growing demand and improved price stability could make it easier for developed countries to overcome domestic resistance to reduced support, thereby facilitating the transition to a more level playing field in world agricultural trade.

NOTES

[1] Although the incidence of poverty is higher in the mineral economies among the LDCs, the relationship between commodity dependence and poverty differs from that in agricultural economies. As a very sweeping generalization, it may be said that while poverty in agricultural LDCs affects those engaged in commodity production, in mineral LDCs it is a problem for those excluded from the production of commodities.

[2] Prices of some manufactured products, for instance, computer chips, are subject to considerable fluctuations. Significantly, the process whereby such products become sufficiently standardized and widely traded for this to happen is usually termed 'commoditization'.

[3] The index is based on a classification of tariffs and non-tariff barriers.

[4] The 'unbound' nature of these preferences is a major shortcoming from the point of view of providing long-term security for investors.

[5] It should be noted that this kind of liberalization is not fully reflected in some developed countries. For example, Canadian and Australian Wheat Boards (the Australian board has been privatized into Australian Wheat Board Ltd.) account for about one third of world wheat exports. The New Zealand Dairy Board handles about 30 % of world dairy exports. (Murphy 1999, pp. 6-7).

REFERENCES

Dehn, J., 2000. *The effects on growth of commodity price uncertainty and shocks.* World Bank, Washington. World Bank Policy Research Working Paper no. 2455.

Murphy, S., 1999. *Market power in agricultural markets: some issues for developing countries.* South Centre, Geneva. South Centre Working Paper no. 6. [http://www.southcentre.org/publications/agric/toc.htm]

Reardon, T. and Berdegué, J.A., 2002. The rapid rise of supermarkets in Latin America: challenges and opportunities for development. *Development Policy Review,* 20 (4), 371-388.

Sachs, J. and Sanchez, P.A., 2004. Une révolution verte pour l'Afrique. *Le Monde* (10 August). [http://www.unmillenniumproject.org/documents/lemonde081004.pdf]

UNCTAD, 1999a. *Impact of changing supply-and-demand market structures on commodity prices and exports of major interest to developing countries.* UN, New York. TD/B/COM.1/EM.10/2. [http://www.unctad.org/en/docs/c1em10d2.en.pdf]

UNCTAD, 1999b. *The Least Developed Countries 1999 report.* UN, New York. [http://www.unctad.org/en/docs/ldc1999_en.pdf]

UNCTAD, 2000. *The Least Developed Countries 2000 report: aid, private capital flows and external debt: the challenge of financing development in the LDCs.* UN, New York. [http://www.unctad.org/en/docs/ldc2000_en.pdf]

UNCTAD, 2002a. *Export diversification, market access and competitiveness.* UN, New York. TD/B/COM.1/54. [http://www.unctad.org/en/docs/c1d54_en.pdf]

UNCTAD, 2002b. *The Least Developed Countries 2002 report: escaping the poverty trap.* UN, New York. [http://www.unctad.org/en/docs/ldc2002_en.pdf]

UNCTAD, 2003. *Background notes prepared for meeting of eminent persons on commodity issues, including the volatility in commodity prices and declining terms of trade and the impact these have on the development efforts of commodity-dependent developing countries.* UN, New York. TD/B/50/CRP.3. [http://www.unctad.org/en/docs/tb50crpd3_en.pdf]

UNCTAD, 2004a. *Development and globalization: facts and figures.* UN, New York. [http://www.unctad.org/en/docs/gdscsir20041_en.pdf]

UNCTAD, 2004b. *The least developed countries report 2004.* UNCTAD, New York. [http://www.unctad.org/en/docs/ldc2004_en.pdf]

UNCTAD and Commonwealth Secretariat, 2001. *Duty and quota free market access for LDCs: an analysis of Quad initiatives.* UN, London. UNCTAD/DITC/TAB/Misc.7. [http://www.unctad.org/en/docs/poditctabm7.en.pdf]

Weatherspoon, D.D., Neven, D., Katjiuongua, H., et al., 2003. *Battle of supermarket chains in Sub-saharan Africa: challenges and opportunities for agrifood suppliers.* UN, Geneva. UNCTAD/DITC/COM/Misc/2003/5. [http://www.unctad.org/en/docs/ditccommisc20035_en.pdf]

WTO, 2004. *World Trade Report 2004: exploring the linkage between the domestic policy environment and international trade.* WTO, Geneva. [http://www.wto.org/English/res_e/booksp_e/anrep_e/world_trade_report04_e.pdf]

CHAPTER 3

MAKING AGRICULTURAL TRADE REFORM WORK FOR THE POOR

M. ANN TUTWILER AND MATTHEW STRAUB

International Food & Agricultural Trade Policy Council, Washington, DC, USA

INTRODUCTION

With almost 800 million people in the developing world suffering from chronic hunger and a total of 1.2 billion people living on less than a dollar a day, poverty remains the greatest failure of the contemporary global economy, as well as the greatest challenge facing the global polity. Past efforts to focus official development assistance and debt relief on poverty alleviation are noble, but it is clear that solutions must look beyond aid to produce meaningful progress. Though foreign aid clearly has a role to play, it cannot be the primary means for sustainable economic development because it alone cannot enlarge the opportunities for economic growth. And without economic growth in developing countries, it is impossible to lift nations, and people, out of poverty.

With agriculture at the heart of current WTO negotiations, much is at stake for the rural poor. There is, however, a great deal of controversy surrounding the role that agricultural trade reforms can play in alleviating poverty and increasing food security. Most economists agree that economies that are more open to trade and investment grow faster than closed economies, and that robust economic growth is the only way to lift people out of poverty. But opponents of trade reforms for developing countries do not agree that open economies grow more rapidly than closed economies, nor do they believe that trade and investment-led economic growth alleviate poverty. Many believe that more open trade and investment regimes exacerbate poverty – particularly in developing countries and especially in agriculture.

This debate is at the centre of the Doha Round trade negotiations. While much of the trade debate has focused on reducing trade-distorting subsidies and improving

N. Koning and P. Pinstrup-Andersen (eds.), Agricultural Trade Liberalization and the Least Developed Countries, 25–50.

market access in developed countries, increasingly developing countries are being asked to open their markets, mainly by lowering their tariff barriers. Developing countries have resisted opening their markets for fear of hurting their farmers. Both developed and developing countries have obtained many exemptions and exceptions (in the form of Sensitive and Special Products) in the market access negotiations. In general, tariff measures are a poor means of helping subsistence farmers. They tend to raise food prices, which harms consumers, and they do nothing to improve farmers' productivity or competitiveness. In many cases, tariffs become a wall for politicians to hide behind, rather than make the investments in rural infrastructure, in communications, in research and extension that will really improve the livelihoods of subsistence farmers.

This paper brings together what is known about the link between agricultural trade reform and poverty alleviation and about how developing countries can successfully manage to open their economies while reducing poverty. It highlights the channels that link agricultural growth, rural development and poverty alleviation with trade. It discusses the potential welfare impacts of policy reform and examines the recent trends of open trade in developing countries. Finally, the paper identifies policy reforms, investments and flanking measures that could be effective in combating poverty in conjunction with a more open trade policy.

AGRICULTURE AND POVERTY

For subsistence farmers, agriculture and the agri-food sector represent the dominant source of potential income and employment. In developing countries, agriculture employs almost three-quarters of the population and accounts for about half of the Gross Domestic Product (GDP). In the poorest of these countries, over three-quarters of the population live on less than two dollars a day – a proportion that is not expected to change and, in some regions, will only grow worse without significant changes in domestic and trade policies (Table 1). Counter-intuitively,

Table 1*. Share of population living on less than $2 per day*

	2001 (%)	Projected for 2008 (%)
East Asia/Pacific	47	32
South Asia	77	72
Eastern Europe/Central Asia	20	21
Latin America/Caribbean	25	39
Middle East/North Africa	23	22
Sub-Saharan Africa	77	77

Source: World Bank: World Development Indicators 2005

subsistence farming cannot solve the most basic needs for food security. While some may assume that such a heavy reliance on farming provides some degree of food security, in fact the opposite holds true. Developing countries deriving a large share of their GDP and employment from the agricultural sector tend to have higher rates of poverty, hunger and malnutrition (Figure 1).

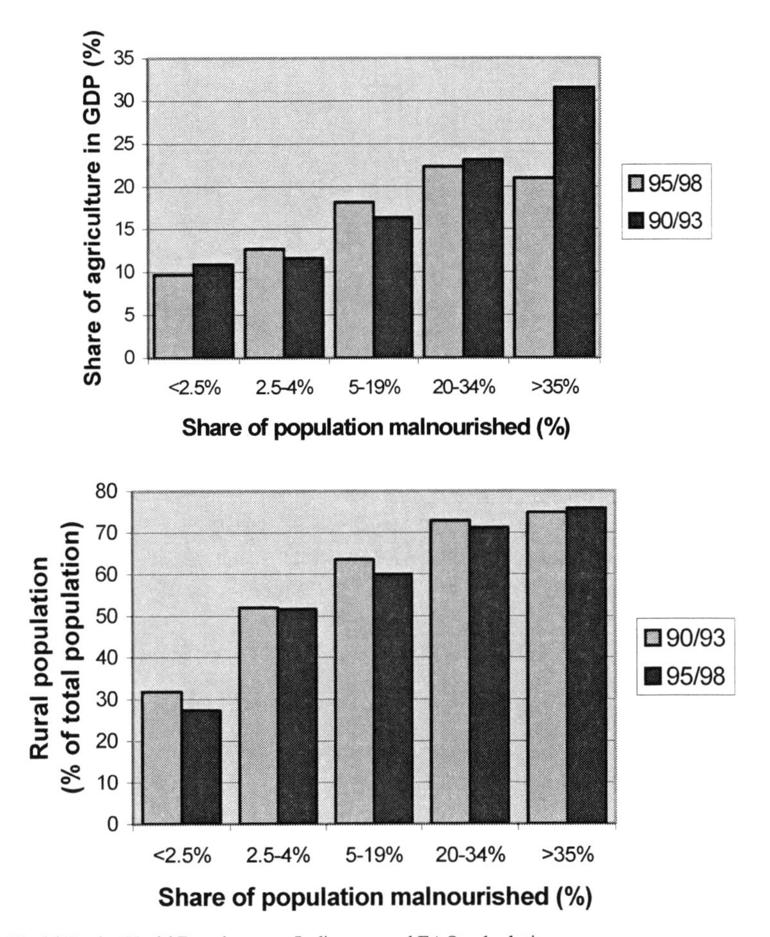

Source: World Bank: World Development Indicators and FAO calculations

***Figure 1**. The importance of the agricultural sector in developing countries by prevalence of the undernourishment category (1990-1993 and 1995-1998)*

TRADE AND ECONOMIC GROWTH

Trade has the potential to lift developing nations out of poverty on a scale that could generate several times any conceivable benefits derived from direct monetary aid. The links between trade and economic growth operate through various channels, including changing the relative prices of tradable goods and the incentives for investment and innovation. Trade acts as a catalyst for economic growth by encouraging investment, efficiently allocating resources and opening markets for those goods that people can produce most competitively. It is important that agriculture participate in this process, because agriculture is the dominant industry in most developing countries; the rural poor make up 75% of the total population in the developing world and suffer the most from deficiencies in capital and technology (FAO et al. 2002).

There is strong evidence that open-trade regimes (and more generally open economies) are associated with higher rates of economic growth. On average, open economies grew 3.5% annually versus closed economies, which grew at less than 1% annually (Berg and Krueger 2003). Over time, the difference in these two growth rates on the level of incomes is stunning: at a one-percent growth rate, it takes 62 years for incomes to double; at 3.5%, incomes will increase 16 times in 62 years. Even a small annual difference in growth rates can be dramatic over the long term.

International trade allows counties to specialize in activities where they hold a comparative advantage. Trade extends the market facing local producers, allowing them to take advantage of economies of scale. Trade encourages more efficient allocation of resources and thereby raises incomes, since finding new and better ways of using land, labour and capital is vital to economic growth.

Openness to international trade is also closely linked to a supportive investment climate (both foreign and domestic), which is positively correlated with economic growth (Table 2). When markets are freed up, private investors see greater opportunity and reduced uncertainty where previous barriers may have restricted their business. Private investment brings intellectual capital and technology, and can also nudge other aspects of social infrastructure in a positive direction. Openness to trade also strengthens the financial services sector, which can better mobilize resources for domestic and foreign direct investment. The effects of trade on investment are often overlooked in models because they involve a more complicated analysis and investor decisions are often difficult to predict. Yet this linkage is vital to the development of a modern economy.

Growth and investment in the agricultural and agri-food sector has an especially important role to play in poverty alleviation because the benefits of increased primary agricultural production spill beyond the sector and spur more general economic growth. First, there is the direct impact of agricultural growth on farm incomes, which account for a large share of the GDP in developing countries. Second, these spin-offs or multiplier effects expand other economic activities because of strong linkages with other sectors. An additional dollar of income in the rural sector generated an additional three dollars in rural income through increased demand for rural goods and services (Watkins 2003). More jobs are created in agricultural-related industries and in the non-farm sector as farmers spend additional income. Third, there are national impacts, including lower prices for food and raw materials to the urban poor, increased savings, and reduced food imports or foreign exchange costs. Therefore, even poor and landless workers who may be net buyers of food benefit from the indirect effects of trade reform through higher wages and an increased demand for unskilled labour.

***Table 2**. Trade, investment, education key sources of economic growth 1960-2000*

	1960-1973		1973-1985		1985-2000	
	Fast growers (%)	Slow growers (%)	Fast growers (%)	Slow growers (%)	Fast growers (%)	Slow growers (%)
Per capita income growth	5	1	3	-1	3	0
Share of investment/ GDP	24	11	21	14	18	12
Ratio of trade/GDP	63	44	72	58	79	64
Primary-school enrollment rate			87	74	98	87
Secondary-school enrollment rate			38	29	60	36

Source: WTO: World Trade Report 2003

In addition, countries that are more open to trade often boast other policies that support macroeconomic stability and development. The agricultural sector in particular requires clear property rights, more research and improved infrastructure to increase competitiveness. These investments are often forthcoming once the economic potential becomes apparent.

ECONOMIC GROWTH AND POVERTY

Economic growth is not an end in itself, but a means to poverty alleviation and the general improvement of people's lives. There is a strong correlation between per capita income and how a country ranks on the Human Development Index (Figure 2)[1]. In fact, many of the outliers (circled in Figure 2) are countries in sub-Saharan Africa, where decreased life expectancy and other health impacts, caused by the AIDS epidemic, have hindered the countries' ability to prosper.

In recent years, improved governance has become a centrepiece of national development strategies. The role of open trade regimes in improving governance is often overlooked. One of the biggest challenges for developing countries is pervasive corruption. Corruption is driven by rent-seeking behaviour, which pervades in the absence of the rule of law and accountable systems of governance. When trade barriers are high, or where import and export quotas are in place, merchants resort to bribery to subvert high tariffs or to buy import and export licenses. Generally, corruption favours the rich and well-connected, who have the means and access to bribe government officials, and hurts the poor, who are often its victims (Berg and Krueger 2003). Corruption is significantly reduced under open trade regimes (Figure 3). There is also a strong correlation between openness to trade and the rule of law (Figure 4).

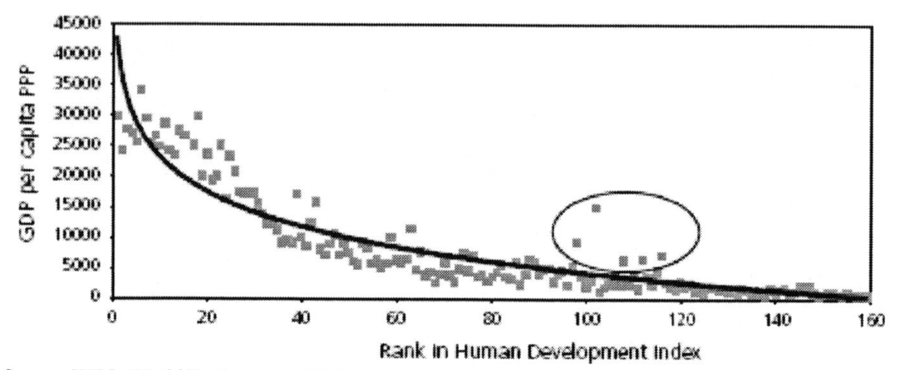

Source: WTO: World Trade Report 2003

***Figure 2**. Higher incomes lead to higher welfare*

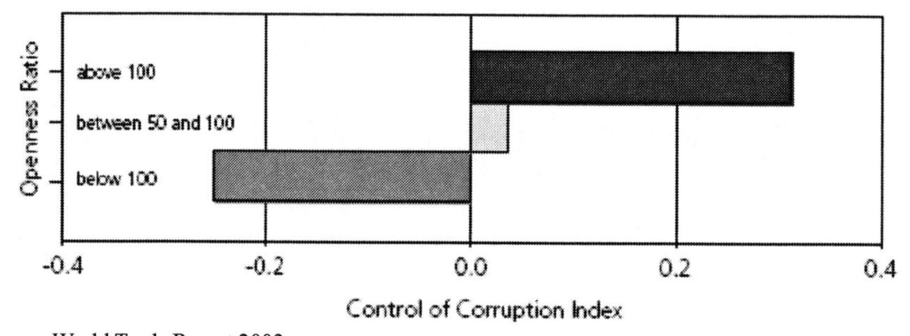

Source: World Trade Report 2003

***Figure 3**. Open economies face reduced corruption*[2]

ECONOMIC GROWTH AND POVERTY ALLEVIATION

Whether economic growth leads to poverty reduction is questioned by many of those who oppose trade reforms. They argue that economic growth does not necessarily improve the lives of the poor. But, a 2000 study of 80 developing countries over the past 40 years demonstrates that the income of the poorest 20% of the population in developing countries increased dollar for dollar with increases in per capita GDP (Dollar and Kraay 2000). Other studies concurred: finding that a one-percent increase in income in developing countries lowers the poverty rate by 1 to 3% (Watkins 2003; Berg and Krueger 2003).

Economic growth can alleviate poverty but not necessarily improve income distribution. China's recent growth is a classic example. There is no question that economic growth has raised many Chinese out of poverty, but it is also true that income distribution in China has worsened. Increasing inequity in income distribution can have adverse consequences. It may, for example, spur unsustainable rural–urban migration, creating social and political problems that may need to be addressed.

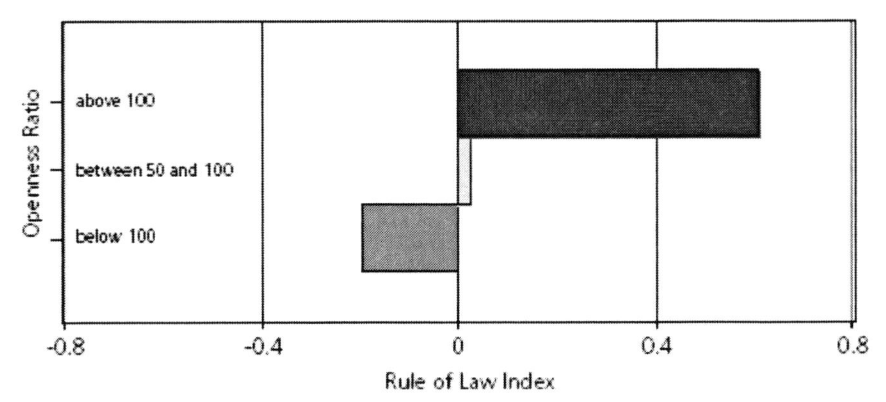

Source: World Trade Report 2003

Figure 4. *.... and greater rule of law*[3]

If income inequality increases, it does not necessarily mean that the poor are absolutely worse off. The poor may be relatively worse off, but absolutely better off as a result of economic growth. For example, trade reform in Bangladesh led to an increase in income inequality, but the percentage of people living below the poverty line fell from 28 to 25%. Similarly, income inequality increased while Chile was opening its market to international trade, but the proportion of people living in poverty fell from 17 to 6% over the span of 16 years (Winters 2002). The case of East Asia is similar. In the mid-1970s, six of every ten people in East Asia lived in extreme poverty. Today, fewer than two in ten live in extreme poverty. The absolute number of people living on less than a dollar a day in the region has fallen from 720 million to 278 million. Average incomes have grown by 5% annually, resulting in a doubling of per capita income every 14 years. This growth was mostly associated with rising exports, which drove the demand for goods in labour intensive manufacturing and generated foreign exchange.

It is important to distinguish between relative and absolute poverty. If poverty is defined as a relative concept, then every country faces some degree of poverty – there are certainly poor people in the United States, but on a relative scale the poor in the US are well-off compared to the poor in Africa or Asia. In the context of development goals, the concept of absolute poverty becomes much more significant because it refers to the ability of families to meet minimum consumption needs, without reference to the income or consumption levels of the general population. The first goal of development should be to reduce the level of absolute poverty.

TRADE AND FOOD SECURITY

The stability of food consumption is a particular concern for developing countries, and for poor households in developing countries. At the national level, the proportion of malnourished people and underweight children tends to be lower in countries where agricultural trade in proportion to agricultural production is large. The fact that farmers currently produce 17% more calories per person today than they did 30 years ago, despite a 70-percent population increase, demonstrates that hunger is a problem of income and access, not global availability (FAO et al. 2002).

Weak access to and integration with international markets limits the ability of countries with widespread hunger to import enough food to compensate for domestic production shortfalls. Figure 5 shows that countries with high incidences of malnourishment import less than 10% of their food, compared to more than 25% in countries with greater food security. According to a 2003 study by the FAO, "The relative isolation from international trade appears to be more a measure of vulnerability than of self-sufficiency" (FAO 2003, p. 18). Access to foreign markets can serve as an insurance policy during production shortfalls, because trade balances domestic production with imports. Trade reduces variability in consumption, as countries are not dependent only on their own highly variable production levels (Diao et al. 2003). (Clearly, increased dependence on trade is partly a function of higher incomes that lead to higher demand for food, and allow consumers to diversify their diets with imported foodstuffs. But, for those who argue that imports are a sign of food insecurity, more often than not the reverse is true.)

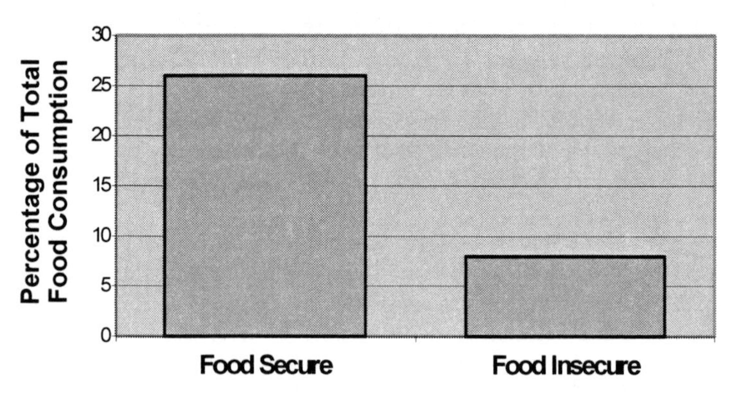

Source: FAO 2003

Figure 5. *Food-secure countries import more than food-insecure countries*

Tariffs and other border measures are often justified on grounds of food security. But tariffs are a poor way to improve food security for society overall (Table 3). While farmers and landowners are better off in terms of access to local food and potentially higher incomes, they are worse off because of higher food prices and limited access to imported food in times of domestic shortfall. (Most subsistence farmers consume more food than they produce, and so are not helped overall by higher commodity prices.) However, the effectiveness of border measures also depends on whether or not the higher prices actually reach poor farmers in rural areas. In many developing countries, the price at the border has virtually no impact in rural areas because of poor infrastructure and poor internal marketing, so tariff measures have a negative impact on urban consumers but little positive impact on subsistence farmers.

Since many of the rural poor are subsistence farmers or altogether landless with little or no surplus crop to sell, the higher prices resulting from tariffs become ineffective when times are good and render them completely dependent on food aid during shortfalls. (Food aid often depresses domestic prices, so that even those local farmers that can manage during the bad times benefit less than they should from protectionist domestic policies.)

Table 3. *Commodity-dependent countries face higher rates of malnutrition*

	Per capita food consumption	Incidence of undernourishment	Probability of consumption shortfall
Single-commodity-dependent exporters	2314	36%	22%
Non-commodity-dependent countries	2285	22%	15%
China	2972	9%	1%
India	2493	24%	8%
All low-income, food-deficit countries	2317	19%	16%

Source: Pingali and Stringer 2003

Lower trade barriers – to both developed and developing country markets – can also help countries diversify their agricultural base. Farmers that have access to developed country and to neighbouring markets can afford to diversify out of staple crops by growing commercial crops. Countries with more diverse agricultural sectors tend to be more food secure. Those countries that depend on a single agricultural export face low food consumption (2300 calories per capita/day), high prevalence of undernourishment (36%) and a relatively high probability of a consumption shortfall (Table 3). Where agricultural sectors are more diversified, there is a lower probability of variations in consumption. As beneficial as preferential access schemes extended by some OECD countries may seem, they have contributed to narrowing the scope of agricultural production for many developing countries. Though such agreements have offered better market access to certain poor countries for specific products, they have done so at the cost of diversified agriculture. Preferential treatment can also mislead producers into growing crops for which they might not be competitive otherwise, at the expense of other crops. Dependency on a single commodity such as sugar or coffee is a particular challenge for several countries in Latin and South America. (There has been a great deal of controversy in the WTO negotiations over the impact of lower developed country sugar tariffs on preference holding countries. Clearly, in the short term, these countries will lose from lower sugar prices in rich country markets. But they will gain from higher world sugar prices, and over time, they will gain as they diversify out of sugar into other crops or other economic activities.)

TRADE NEGOTIATIONS

The Doha Development Round of WTO negotiations has centred on the need for developed countries to reduce trade distorting domestic subsidies; phase out subsidized export competition; and open up markets by reducing tariffs and increasing import quotas. The extent of trade reform required of developing countries has received far less attention. Unlike the developed countries, developing countries rely almost exclusively on tariffs to protect their domestic markets. Few developing countries have the financial resources to offer direct subsidies to their farmers.

Numerous economic studies have examined the economic benefits of trade reform and, while different methodologies have produced a range of resulting effects, the direction of these effects has consistently shown overall welfare gains from trade. Estimates for the total economic gains from eliminating OECD agricultural protection range from $8 billion to $26 billion. Estimates of the benefits for developing countries from multilateral reform range from $2.6 billion to $21.5 billion (Beierle and Diaz-Bonilla 2003).

Table 4. *Rich-country trade reforms increase income to developing countries (increase in millions of US dollars)*

Changes in agricultural trade policies by:	United States only	European Union only	Japan/Korea only	All industrialized countries
Sub-Saharan Africa	$455	$1,290	$150	$1,945
Asia	$2,186	$2,099	$2,346	$6,624
Latin America and the Caribbean	$2,896	$4,480	$607	$8,258
Other developing countries	$1,148	$5,069	$339	$6,659
All developing countries	$6,684	$12,936	$3442	$23,486

Source: IFPRI 2003

Most economic studies also find that the impact from multilateral trade reform on developing country welfare and food security is positive. For example, a study by the International Food Policy Research Institute (IFPRI), released just prior to the Cancun Ministerial, showed significant gains to developing countries from the complete elimination of trade distorting subsidies and tariff barriers in developed countries (Table 4)[4].

According to the IFPRI report, global benefit to all developing countries was $23.4 billion, with Latin American and Asian countries gaining the most and even sub-Saharan Africa gaining almost $2 billion in additional income (Diao et al.

2003). Table 5 details these results for select countries as a relative annual increase in income.

***Table 5**. Rich country trade reforms increase income to farmers and the agricultural sector (percentage increase)*

	Removal of all OECD subsidies and border protection
Argentina	3
Brazil	3
China	2
India	1
East Asia (rest)	1
Latin America (rest)	8 (1-15% range)
Sub-Saharan Africa	9 (3-15% range)

Source: IFPRI 2003

A study completed for the World Bank that includes dynamic benefits due to increased domestic/foreign investment and innovation illustrates that income would increase up to 80% in Argentina and 57% in sub-Saharan Africa over 15 years (Table 6).

***Table 6**. Multilateral trade reforms increase incomes to developing countries (percent change from baseline in 2015)*

Countries/Policies	**Removal of all subsidies and border protection**	**Removal of border protection only**
Argentina	80	44
Brazil	32	17
China	-4	7
India	23	16
East Asia (rest)	7	6
Latin America (rest)	72	65
Sub-Saharan Africa	57	52

Source: Beghin et al. 2002

Whether reductions in trade distorting subsidies will benefit the poorest developing countries depends on the composition of their agricultural production. For example, 90% of Africa's exports are in ten commodities (cocoa, coffee, cotton, tobacco, sugar, tea, palm oil, rubber, bananas and peanuts) (Beierle and Diaz-Bonilla 2003). Some of these products compete with subsidized temperate zone products on export markets, but many do not. Reducing distortions in products such as sugar, peanuts tobacco and cotton, which are highly protected in OECD markets, would benefit African exporters. (Although, as noted above, some preference holders would lose from sugar policy reforms in the short term, but would gain over the longer term.) In other cases, African smallholders compete with subsidized imports, such as maize, which are dumped into their domestic markets. OECD policies tend to depress prices in the range of 5 to 20%, thereby allowing producers to 'dump' goods onto the international market. In such cases, improved trade rules would help African farmers who cannot compete with these products in the domestic market. The removal of trade distorting OECD subsidies that encourage dumping would tend to boost agricultural prices, substantially in some cases.

Accordingly, most economic studies predict that commodity prices will also rise following multilateral reform. (However, even if trade reforms raise commodity prices above the subsidy-depressed levels prevalent today, trade reform would not reverse the long-term decline of commodity prices.) Price increases are expected to be sharper for commodities that low-income net-food-importing countries must import, than for the commodities they export (FAO 2003). Price increases will therefore tend to help commercial farmers and exporting countries, and tend to disadvantage subsistence farmers and importing countries, depending on what the country produces and consumes, how domestic prices change, and whether the country is a net importer or exporter.

Subsidies benefit net food importers in the short term by depressing prices of food imports. By raising world market prices, subsidy cuts will increase the price of imported staples (rice, wheat and other grains) that are heavily subsidized by OECD countries. However, these price increases are expected to be relatively small. An IMF study predicts that removing all policy distortions would increase the price of seven out of ten commodities by less than 4%, with large increases for the other three: milk (24%), refined sugar (8%) and sheep meat (22%) (Tokarick 2003). Removal of production and input subsidies alone increase prices by less than 2% (Table 7). And, because subsidies will be phased out over a long time frame, the increase in prices due to trade reform will likely be overshadowed by annual fluctuations in demand and supply.

Least Developed Countries are more likely to benefit from subsidy cuts than other developing countries. On average, 18% of their exports are subsidized by at least one OECD member (compared with 4% of exports by other higher-income developing countries), meaning LDCs are competing against OECD governments more often than highly developed countries. For example, Benin, Burkina Faso, Chad, Malawi, Mali, Rwanda, Sudan, Tanzania, Uganda and Zimbabwe have 60 to 80% of their total exports subsidized by one or more WTO members. On the other hand, 9% of LDC imports involve products subsidized by OECD countries, compared to only 4% of the imports of other developing countries (Hoekman et al. 2002).

***Table** 7. Commodity prices increase as a result of trade reforms (percent change from baseline in 2015)*

Commodities/Policies	Removal of all subsidies and border protection	Removal of border protection only
Paddy rice	6	4
Wheat	12	2
Horticultural products	0	0
Oilseeds	8	1
Refined sugar	9	8
Beef	10	2
Dairy products	8	6

Source: Beghin et al. 2002

Counter to most press coverage of the Doha Development Round, many studies indicate that for most developing countries, reducing developed and developing country tariffs is more important than reducing developed country domestic subsidies. Reducing tariffs generates significantly higher benefits in these studies than lowering subsidies, in part because both developed and developing countries rely on tariffs, while only developed countries use subsidies. Therefore, on a global scale, markets are more distorted by tariffs than by subsidies – and removing those distortions will generate across the board economic benefits. Moreover, even though developed countries have far higher subsidies than developing countries, developing countries have higher tariffs than developed countries. (This is not to argue that reducing trade distorting subsidies is not important. These subsidies do depress global prices, and they provide a rationale for many developing countries to maintain high tariffs against subsidized competition. For both political and economic reasons, these trade distorting subsidies should be reduced, and ultimately phased out.)

Studies by the OECD and the IMF support this conclusion: the impact of tariffs is much more deleterious than the impact of subsidies on developing countries. An IMF study cautions that scraping subsidies without complementary cuts to tariffs would help big developing exporters such as Brazil, Argentina and, to some extent, China and India, but would harm the rest of the developing world. It estimates that South Asia would be $164 million worse, sub-Saharan Africa would suffer a $420 million loss and North Africa and the Middle East would lose the most at $2.9 billion. By contrast, the same study predicts that eliminating all developed country tariffs on agricultural imports would produce a total welfare gain of $91 billion and no countries would be harmed (Tokarick 2003). Table 8 illustrates the results of various studies, supporting the IMF's findings and indicating a more general consensus that removing OECD tariffs by themselves boosts the incomes of LDCs.

***Table 8**. All studies show developing countries gain from reform*

Authors	Reforms analysed	Global effect ($US billion)	Developing countries ($US billion)
Diao et al. (2001)	(i) Removing all agricultural supports and protections	31	3
	(ii) Removing all tariffs	25	6
	(iii) Removing domestic supports in the developed countries	3	-2
	(iv) Removing export subsidies, worldwide	0	-2
Hertel et al. (1999)	(i) 40% reduction in all agricultural protection	70	15
	(ii) Same excluding production subsidies	60	15
Francois et al. (2003)	(i) Full liberalization of border measures	97	25
	(ii) Liberalization of OECD border measures	39	4
	(iii) Liberalization of non-OECD border measures	59	21
	(iv) Full liberalization of OECD domestic support	12	-2
Anderson et al. 2000)	(i) Full liberalization of all protection	165	43
	(ii) Full liberalization of all protection in OECD	122	31
	(iii) Full liberalization of all protection in non-OECD	43	12
Brown et al. (2001)	33% reduction in post-Uruguay protection of agriculture	-3	-16
Dee and Hanslow (2000)	Elimination of all post-Uruguay trade barriers in agriculture	50	
UNCTAD (2003)	(i) 50% cut in all agricultural tariffs	22	10
	(ii) Elimination of export subsidies in agriculture	-2	-6
	(iii) Tariffs are reduced by 50% on processed agriculture	12	6
Dimaranan et al. (2003)	(i) 50% reduction in OECD domestic support	---	-0.4
Hoekman et al. (2002)	(i) 50% cut in all agricultural tariffs	14	2
	(ii) 50% cut in domestic support	0.3	-0.2

Source: Charlton and Stiglitz 2005

While the prevailing wisdom has been that developed country reforms are good for developing countries, and developing country liberalization is good for developed countries, the opposite is in fact truer: each group of countries benefits most from its own liberalization, not from liberalization by the other group. A recent study demonstrates that developing countries benefit even more from reducing their own tariffs than from a reduction in OECD tariffs, because high tariffs raise food prices to consumers and limit opportunities for other developing countries (Table 9) (Anderson 2004). Another study reinforces this point, arguing that anticipated Doha reforms are less poverty-friendly than full liberalization because they do not envisage meaningful cuts in developing country tariffs (Hertel and Winters 2005). The lion's share of the gains accrue to the liberalizers themselves, and if developing countries want to maximize their benefits from trade negotiations, their first best solution is to free up their own agricultural markets, mainly by cutting tariffs. About three-quarters of the benefits that come from developing country trade reform will stay with them, while only 10% of the benefits from developed country reforms would accrue to developing countries (Table 9). These results reflect the importance of own country reform and the expansion of South-South trade – trade among developing countries – in the overall scheme of trade liberalization.

Table 9*. Low-income and developed countries benefit from their own reforms distribution of gains from full trade liberalization (in 1995 US$ billion)*

Liberalizing region		**Agriculture and food**	**Other primary**	**Textiles and clothing**	**Other manufactures**	**Total**
Region	*Benefiting region*					
High income						
	High income	110.5 (90%)	0.0	-5.7	-8.1	96.6
	Low income	11.6 (10%)	0.1	9.0	22.3	43.1
	Total	122.1 (100%)	0.0	3.3	14.2	139.7
Low income						
	High income	11.2 (26%)	0.2	10.5	27.7	49.6
	Low income	31.4 (74%)	2.5	3.6	27.6	65.1
	Total	42.6 (100%)	2.7	14.1	55.3	114.7
All countries						
	High income	121.7 (74%)	0.1	4.8	19.6	146.2
	Low income	43.0 (26%)	2.7	12.6	49.9	108.1
	Total	164.7 (100%)	2.8	17.4	69.5	254.3

Source: Anderson, 2004

IMF and OECD analyses confirm these results. Although the dollar gains are smaller in absolute terms when compared to the gains that accrue to rich countries, they are far greater as a percentage of developing-country GDP. Anderson puts these gains at 1.9% of GDP for developing countries, over three times the expected gains for rich countries (2004).

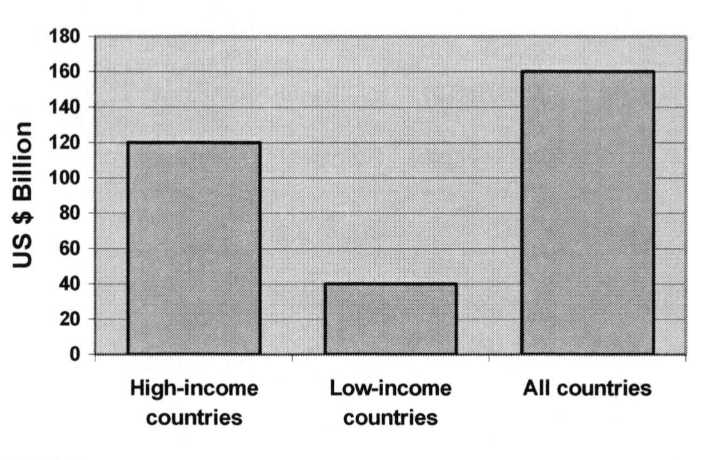

Source: OECD, 2003

***Figure 6**. Developed, developing countries gain from agricultural trade reform*

High tariff and non-tariff barriers between developing countries inhibit the potential for South-South trade, and thus the ability for developing countries to help each other. South-South trade can reduce the dependence of the South on the North; to encourage diversification of production in the South; and to capitalize on geographic proximity. Goods produced in the South may also be better suited for neighbouring markets with similar income levels, tastes, cultures and regulatory systems. In terms of improved productivity, South-South trade offers opportunity for developing countries to introduce new technologies and resources through other neighbouring developing countries where FDI is steadily rising, thereby offering each other mutual support. (Reducing tariffs in developing countries across the board is more beneficial than creating regional customs unions in developing countries which result in trade diversion. However, along with cutting tariffs, South-South trade will need to be accompanied by reforms in custom procedures and regulatory measures.

A final argument against reducing tariffs in developing counties is that developing countries rely on tariffs to generate revenues. In many developing countries, tariff revenues comprise a significant share of government resources. Twenty-five developing countries derive 30% of their total tax revenue from tariffs, according to an IMF study. Lowering these tariffs may deprive governments of much needed funds. But, if tariffs are reduced, overall trade volumes are likely to expand, potentially cancelling out the reduction in tariff levels. The increased growth generated by more open trade may provide governments with higher net tariff revenues. Also, high tariffs drive trade underground onto the black or the grey market, something that is already a problem in many developing countries (Bannister and Thugge 2001). Lowering these tariffs may in fact bring some trade from the 'informal' back to the formal sector. Third, there are no guarantees that these government revenues are used to benefit rural sectors, and fourth, high tariffs may contribute to corruption as traders seek to avoid paying these taxes.

TRADE REFORM AND DOMESTIC POLICY

Since the poor in most developing countries are subsistence farmers who live in rural areas, it is vital that trade reforms be coupled with domestic policy reforms and financial assistance to ease the transition and facilitate the adjustment. Trade exerts a positive impact on economic growth because it serves as conduit for new ideas, new technology, and competition, which drives innovation and increases productivity. However, if markets are missing or do not function properly, then shifts in relative prices (i.e. from tariff cuts) will not lead to a shift of production, jobs and investment. Similarly, the lack of good roads, ports, telecommunications and marketing infrastructure can hamper a country's ability to participate in and benefit from international trade. Some of these conditions are inherited from geography; some are the result of inadequate or misguided investments, and others are relics of colonial rule.

For poor people to benefit from reform, they must be able to participate in markets. Policies that enhance their ability to participate in the formal market will ease the transition to reform. Where markets have been liberalized without accompanying policy reforms and investments (such as improved roads, improved communications, marketing infrastructure), the impact on food security and poverty has been detrimental. Complementary policies must also ensure that reform has a positive effect on people living in rural areas, not just urban centres or favoured areas. The links between poverty and other national policies in education, health, land reforms, micro-credit, infrastructure and governance are as important as border measures.

Trade is not a cure-all for poverty and slow growth, and to some extent it is unfortunate that the name Doha 'Development' Round has led to overly ambitious expectations about the results of trade reform alone. As the WTO's 2003 World Trade Report acknowledges, open trade must be part of a constellation of policies that are pro-growth; macroeconomic stability, reliable infrastructure, transparency, predictability, functioning domestic markets, and a good investment climate all advance the gains of economic efficiency from reduced trade barriers. The common thread behind these reforms is creating flexibility in factor markets – labour, land and capital – that allows the economy to grow.

Labour. The flexibility of labour markets is critical to how trade reform will affect poverty. If firms cannot adjust their work force because of labour regulations, then most of the adjustment to trade will come in the form of changes in wages to already employed workers. If workers can move from one sector to another, or if there is a minimum wage in place, then the adjustment will come in the form of changes in the level of employment.

For example, there are 749 products that are reserved for small-scale firms in India. Other tax exemptions and production subsidies also favour small-scale producers. These policies prevent firms from growing and ultimately from competing in the export market. While in the short term such regulations may keep excess labour employed (in fact, Indian firms say they have 17% more labour than they need), over the longer term these regulations inhibit economic growth and employment generation (Stern 2002). Other evidence suggests that the entry rates of firms into liberalized sectors of the economy are 20% higher than into closed sectors of the economy, leading to higher job creation (Berg and Krueger 2003).

In rural areas of most developing countries, the labour market is very flexible and highly responsive, although unskilled. For the rural poor, adjustments to trade reform will take place largely through changes in employment levels. Trade reforms that turn the terms of trade against agriculture will lead to higher unemployment in rural areas. If trade reform improves the agricultural terms of trade, then the likely result will be higher rural employment, benefiting the poor. For example, reforms in East Asia improved the terms of trade for farmers, while reforms in Central Europe turned the terms of trade against farmers. As a result, agricultural output rose in Asia, and it fell in Central Europe, with predictable consequences for the rural sector. (It should be pointed out that, prior to undertaking these reforms, in neither East Asia nor in Central Europe did agricultural prices or input prices reflect the market. The direction of reforms in both sets of countries moved the market towards a more 'rational' allocation of resources in the rural sector.)

Countries such as China and Taiwan realized the need for institutions that enhance the productivity of rural labour, and that the process of poverty alleviation begins in agriculture. But, they also recognized that agriculture is not a long-term solution. Policies must accommodate demographic shifts; particularly in easing the rural–urban transition. There is a high correlation between education and poverty alleviation because education can go a long way in easing movements between rural and urban areas and from agriculture to other sectors. In general, the best way to improve conditions for the poor continually is to expand their opportunities for productive and remunerative employment, including schemes that promote entrepreneurship and innovation.

Land. Initial land distribution and economic structure influences the impact of trade reform on the different groups in society. For example, declining growth rates in South America, coupled with uneven land distribution, has particularly hurt the poor. And, while agriculture should have played a bigger role in these countries' economic growth strategies, it has generally been ignored or discriminated against.

Property-rights reform gave strong income and asset control to producers in Asia and Central Europe. For example, in East Asia, governments deliberately provided incentives to farmers through property rights reform. At the same time, governments also restructured farms to more efficient sizes. In Central Europe, reforms gave land back to farmers who had lost it during collectivization. By contrast, land reforms in the former Soviet Union have been more gradual, less clear and less effective. A more recent focus on gender policy recognizes that granting land rights and enabling women to manage and control their families' assets and wealth produces a more effective development strategy.

The experience of these 'transition economies' varied depending on whether they pursued a gradual or a rapid approach to land reform and privatization of markets. It is also important to note that neither full privatization of property rights nor fully private markets were necessary in either Asia or Central Europe, as long as the rights were perceived as strong and enduring and the hybrid markets reflected a real market.

Reforms in the rest of the economy. Another factor determining how an economy weathers trade reform depends on how widely the economy is reformed. Transition costs are lower when reform occurs across a wide swath of the economy. This is why it is vital that the Doha Development Round include reform in non-agricultural market access and services. Transition costs are lower when governments make accommodating domestic reforms. For example, if monopolies dominate a sector (particularly agricultural suppliers or purchasers) or if price controls limit adjustment, or if labour markets are inflexible, removing these constraints widens the domain of trade and eases adjustments.

Successful countries have recognized that increased rural productivity and development through agriculture will prime the pump for wider economic growth. But, long-term policy solutions must acknowledge the eventual need for a more diversified economy. This requires an investment in human capital (education, training, health care, etc.) and anticipating shifts from primary industries that utilize the land and natural resources, to secondary industries such as manufacturing and processing, and finally to the service-oriented sectors. This has been the path of most modern societies (Canada, Australia, Britain and the U.S.), and is illustrated by several works in progress throughout Asia and Eastern Europe.

The emergence of institutions of exchange is crucial to the success of trade reforms. For example, the countries of the former Soviet Union, by rapidly removing the centrally planned institutions of exchange before market structures were in place, created widespread short-term disruptions in agricultural production. The East-Asian economies gradually replaced the planned economy with a more market-oriented system, beginning instead by gradually raising the prices paid to farmers for their crops.

While developing country resources are limited, many of the reforms required involve legal and institutional changes, not infusions of money. And in any event, all governments make choices on how they distribute available resources between different groups in society. Developing country governments can leverage international aid toward the goal of poverty alleviation by seeking funding for infrastructure and other investments that will connect the poor to markets. If poverty alleviation is an important goal, then efforts can begin immediately by channelling existing resources, including donor aid and World Bank investments, into promoting the institutions that surround agriculture and agricultural research in developing countries.

CASE STUDIES IN TRADE AND DOMESTIC REFORM

In the early 1990s, Mozambique removed a ban on raw cashew exports, which was originally imposed to guarantee a source of raw nuts to its local processing industry and to prevent a drop in exports of processed nuts. As a result, a million cashew farmers received higher prices for the nuts in the domestic market. But, at least half the higher prices received for exports of these nuts went to traders and not to farmers, so there was no increase in production in response to the higher prices. At the same time, Mozambique's nut processing industry lost its guaranteed supply of raw nuts and was forced to shut plants and lay off 7000 workers (FAO 2003). A gradual removal of the ban would have allowed local producers to compete better in the international market, claim higher margins on their raw cashews and thereby encourage more production.

In Zambia, before reform, maize producers benefited from subsidies to the mining sector, which lowered the price of inputs such as fertilizer. A state buyer further subsidized fertilizer for small farmers. When these subsidies were removed and the parastatal privatized, larger farmers close to international markets saw few changes, but small farmers in remote areas were left without a formal market for their maize. In this case, decreasing subsidies over time and investing in research to reduce fertilizer applications would have afforded smaller farmers the means to adapt.

In Vietnam, on the other hand, trade reform was accompanied by tax reductions, land reforms, and marketing reforms that allowed farmers to benefit from increased sales to the market. As Vietnam made these investments, they began to phase out domestic subsidies and to reduce border protection against imports. An aggressive program of targeted rural investments accompanied these reforms. During this reform, Vietnam's overall economy grew at 7% annually, agricultural output grew by 6%, and the share of undernourished people fell from 27 to 19% of the population. Vietnam moved from a net importer of food to a net exporter (FAO 2003).

Similarly, in Zimbabwe, before reform of the cotton sector, the government was the single buyer of cotton from farmers, offering low prices to subsidized textile firms. Facing lower prices, commercial farmers diversified into other crops (tobacco, horticulture), but smaller farmers who could not diversify suffered. Internal reform eliminated price controls and privatized the marketing board. The result was higher cotton prices and competition among three principle buyers. Poorer farmers benefited through increased market opportunities, as well as better extension and services. As a result, agricultural employment rose by 40%, with production of traditional and non-traditional crops increasing.

Such efforts must also bring developing countries up to speed in meeting current international standards for productivity and quality. One of the main challenges for developing countries to open their markets will come in the form of technical standards, and bringing small farmers in line with the requirements of the Sanitary and Phytosanitary Standards (SPS) and the Technical Barriers to Trade (TBT) – among other treaties – will require technical assistance through aid.

Initial conditions. Initial conditions determine how the changes in relative terms of trade will affect agriculture when reform takes place. Three sets of initial conditions affect the output and productivity in the transition toward reform. Differences in initial price distortions; differences in technology that affected farm restructuring; and differences in the agri-food chain that distorted markets or pseudo-markets are the three main constraints. For example, in East Asia, pro-urban policies used low agricultural procurement prices to subsidize consumers. Price reforms that moved toward more realistic valuations raised prices to farmers, but disadvantaged urban consumers. In Central and Eastern Europe and the former Soviet Union, agricultural prices had been supported at above 'market' levels, and inputs were heavily subsidized. So, price reform in those countries caused substantial declines in the agricultural terms of trade.

In Africa, where there are huge disparities between urban and rural areas left over from colonial times, trade reforms that boosted export crops reversed rural declines in some countries. Where these crops were grown by smallholders the impact on rural welfare was positive. However, the benefits were lower than they might have been, because many countries waited to make reforms until the situation was fairly desperate. Government services had deteriorated and external debt had increased. Countries that succeeded usually did so after prolonged civil strife, when citizens were eager for some change, and trusted charismatic leadership that could lead countries through difficult transitions.

In Latin America, the hugely uneven distribution of land, capital and education has skewed income growth in favour of those with land. This uneven distribution of land and social capital has been exacerbated by growth policies (trade and others) that have favoured large landholders and more highly educated labour.

AGRICULTURAL TRADE REFORMS: NOW OR LATER?

Some policy makers argue that while trade reform may be good for the economy, some sectors may not be able to withstand the competition. These sectors should be 'protected' from competition until they are stronger. There are several problems with this line of reasoning. First, it is impossible to identify, in advance, which sectors should be protected and which can ultimately survive. Second, and more compelling, protection begets protection once the political forces are lined up to lobby and support it. As experience in the United States, Japan and Europe has proven, it is difficult to dismantle protections and subsidies once political forces have captured them.

Foregoing reform altogether or simply postponing it in those sectors that may have a greater impact on the poor is not the solution, either. The evidence suggests that in the long run, this would hurt the poor further by perpetuating slow growth and distorting incentives for investment and innovation in the economy. In any case, as shown above, trade policy is not a very transparent or efficient policy to use to maintain incomes. There are better policy alternatives, even in developing countries, to help maintain poor people's incomes. Moreover, adjustment costs are not usually large in relation to entire economy – and are usually small relative to the benefits.

Alternatively, policy makers argue to postpone reform until a better time. This may be true if the country is in the midst of a recession, when the pain of adjustment is likely to be magnified and the impacts on the least fortunate more difficult to ameliorate. However, there is a difference between trade reform with a long adjustment period (such as that provided by Special and Differential Treatment) and postponing reform altogether. Interestingly, sometimes reform moves faster than scheduled, even when long time frames are envisioned in trade agreements. There is a downside of long transition periods: elected governments may be tempted to push off necessary reforms and adjustments to the next election and perhaps to another party, leaving the protected sector further and further behind. This occurred in the NAFTA agreement in sugar (in the U.S.) and maize (in Mexico), where neither country put in place the needed reforms until the last minute. Then the needed reforms looked too large and too painful to be politically palatable. Moreover, those who wait risk being left behind by other countries seeking to expand their market opportunities.

TRADE REFORM CREATES WINNERS AND LOSERS

Regardless of trends, levels of hunger and poverty differ widely even among countries with very similar levels of agricultural trade. This indicates that the impact of agricultural trade on food security is mediated by other factors, and highlights the fact that trade reform must be accompanied by other policy reforms and investments if it is to have a positive impact on food security. Targeted social measures and investments in rural infrastructure are two measures that enhance food security for farmers and landowners without undermining the whole society's food security (FAO 2003).

Often employment decreases in the short run, but increases over the longer run or in different economic sectors. In general, the employment transition to reform is small relative to the overall size of the economy and the natural dynamics of the labour market. But, even though there are long-term and economy-wide benefits to trade reform, there may be short-term disruptions and economic shocks that may be hard for the poor to endure. Ultimately, the poor may find better jobs in another sector, but weathering that transition may be difficult for those with few resources.

Countries must consider the impacts of trade reforms on the poorest members of society, and formulate policies to counterbalance reforms that adversely affect the poor. Once a government decides to undertake a reform, the focus should be on easing the impact of reforms on the losers – through reforms in labour, land and capital markets, and through education, retraining and income assistance. Government policy should also focus on helping those who will be able to compete in the new environment take advantage of new opportunities.

Notwithstanding the overall positive analyses of the impact of trade reforms on developing countries, economic studies do not always address the significant variations by country, commodity and different sectors within a developing country. Most of the models aggregate all but the largest developing countries into regional groupings, so it is difficult to determine the precise impacts on individual countries.

Even those studies that show long-term or eventual gains for rural households or for the poor, in general do not focus on the costs imposed during the transition from one regime to another.

It is particularly difficult to evaluate the micro-level impacts on different types of producers within different countries, such as smallholders and subsistence farmers. Households with some food surpluses might benefit while subsistence farmers may remain unaffected. Empirical analysis is also not effective at evaluating how those policies will affect poverty among different households or among women and children within households. For this reason, the within-country distributional issues deserve more scrutiny.

Even though trade promotes economic growth and alleviates poverty, it is still important to pursue trade reform with a pro-poor strategy. In other words, focus on reforming those sectors that will absorb unskilled labour from rural areas as agriculture becomes more competitive, and on trade reforms in economic sectors that employ people in deprived areas. Alan Winters has proposed a useful set of questions that policy makers should ask as they consider trade and accommodating reforms (2002):

1) Will the effects of changed border prices be passed through the economy? If not, the effects – positive or negative – on poverty will be muted.
2) Is reform likely to destroy or create markets? Will it allow poor consumers to buy or sell new goods?
3) Are reforms likely to affect different household members – women, children – differently?
4) Will spillovers be concentrated on areas/activities that are relevant to the poor?
5) What factors – land, labour and capital – are used in what sectors? How responsive is the supply of those factors to changes in prices? How flexible is the market?
6) Will reforms reduce or increase government revenue? By how much?
7) Will reforms allow people to combine their domestic and international activities, or will they require them to switch from one to another?
8) Does the reform depend on or affect the ability of poor people to assume risks?
9) Will reforms cause major shocks for certain regions within the country?
10) Will transitional unemployment be concentrated among the poor?

CONCLUSIONS

All policies create winners and losers, including the existing policy environment in developed and developing countries. The losers from the current policy framework – with agricultural trade badly distorted by rich countries' subsidies, markets closed by rich and poor country barriers and insufficient attention to the rural poor in developing countries – are the hungry and the poor. The evidence is consistent and overwhelming that reducing distortions in agriculture, increasing market access and *at the same time* creating a domestic policy environment that supports agricultural and rural areas will increase economic growth and alleviate poverty.

The Millennium Development Goals and the Doha Development Round commit countries to reforming their trade and domestic policies in agriculture. This commitment has been made by OECD countries, which maintain high levels of agricultural subsidies and protection against commodities that are vital to the economic well-being of developing countries. The OECD countries must reduce their trade barriers and reduce and reform their domestic subsidies, but if developing countries are to derive benefits from trade reform, they must reform their trade and domestic policies as well. With the long implementation periods foreseen in the Doha Development Round, developing countries have at least 15 years to make those reforms and investments to enable them to take advantage of trade opportunities and to ease the transition.

Open trade is one of the strongest forces for economic development and growth. Developing countries and civil society groups who oppose these trade reforms in order to 'protect' subsistence farmers are doing these farmers a disservice. But, developing countries and civil society are correct that markets cannot solve every problem, and that there is a vital role for government, for public policies and financial aid. As the Doha negotiators move toward the discussion of modalities, the energies of the international community, developing countries and civil society would be better used to ensure that developing countries begin to prepare for a more open trade regime by enacting policies that promote overall economic growth and promote agricultural development. Their energies would be better spent convincing the population (taxpayers and consumers) in developed countries of the need for agricultural trade reform, and convincing the multilateral aid agencies to help developing countries invest in public goods and public policies to ensure that trade policy reforms are pro-poor.

Trade reform, by it self, does not exacerbate poverty in developing countries. Rather, the failure to alleviate poverty lies in the underlying economic structures, adverse domestic policies, and the lack of strong flanking measures in developing countries. To ensure that trade reform is pro-poor, the key is not to seek additional exemptions from trade disciplines for developing countries, but to ensure that the WTO agreement is strong and effective in disciplining subsidies and reducing barriers to trade by all countries.

Open trade is a key determinant of economic growth, and economic growth is the only path to poverty alleviation. This is equally true in agriculture as in other sectors of the economy. In most cases, trade reforms in agriculture will benefit the poor in developing countries. In cases where the impact of trade reforms is ambiguous or negative, the answer is not to postpone trade reform. Rather, trade reforms must be accompanied by flanking policies that make needed investments or that provide needed compensation, so that trade-led growth can benefit the poor.

NOTES

[1] The Human Development Index measures a country's performance in relation to health, education and income.

[2] The Control of Corruption Index can take values between -2.5 and 2.5 and has been averaged across countries grouped according to their level of openness, where openness is calculated as exports plus

imports divided by GDP. The sample includes 187 countries; 54 countries fall into the range of most open economies, 48 in the range of least open economies and 84 countries in the intermediate range.

[3] As with Figure 4, the Rule of Law Index can take values between -2.5 and 2.5 and has been averaged across countries grouped according to their level of openness.

[4] The model IFPRI used counts all the longer-term, dynamic benefits that would be expected from less distorted markets.

SOURCES AND REFERENCES

Anderson, K., 2004. *Agricultural trade reform and poverty reduction in developing countries.* World Bank, Washington. World Bank Policy Research Working Paper no. 3396.

Bannister, G.J. and Thugge, K., 2001. *International trade and poverty alleviation.* International Monetary Fund, Washington. IMF Working Paper no. WP/01/54. [http://www.imf.org/external/pubs/ft/wp/2001/wp0154.pdf]

Beghin, J.C., Roland-Holst, D. and Van der Mensbrugghe, D., 2002. *Global agricultural trade and the Doha Round: what are the implications for North and South? Presented at the OECD-World Bank Forum on Agricultural Trade Reform, Adjustment, and Poverty, Paris, May 23-24, 2002 and at the Fifth Conference on Global Economic Analysis, Taipei, June 5-7, 2002.* Iowa State University, Ames. Center for Agricultural and Rural Development, Iowa State University Working Paper no. 02-WP 308. [http://www.card.iastate.edu/publications/DBS/PDFFiles/02wp308.pdf]

Beierle, T.C. and Diaz-Bonilla, E., 2003. *The impact of agricultural trade liberalization on the rural poor: an overview: paper presented at Resources for the Future workshop on Agricultural Trade Liberalization, November 3-4, 2003.* IFPRI, Washington.

Ben-David, D., Nordström, H. and Winters, A.L., 2000. *Trade, income disparity, and poverty.* WTO, Geneva. World Trade Organization Special Study no. 5.

Berg, A. and Krueger, A.O., 2003. *Trade, growth, and poverty: a selective survey.* International Monetary Fund, Washington. IMF Working Paper no. 03/30.

Charlton, A.H. and Stiglitz, J.E., 2005. A development-friendly prioritisation of Doha Round proposals. *The World Economy,* 28 (3), 293-312.

Diao, X., Diaz-Bonilla, E. and Robinson, S., 2003. *How much does it hurt? The impact of agricultural trade policies on developing countries.* IFPRI, Washington. [http://www.ifpri.org/media/trade/trade.pdf]

Dollar, D. and Kraay, A., 2000. *Growth, trade reform, and poverty: a macroeconomic approach: paper presented at the conference Poverty and the International Economy, Stockholm, October 20-21, 2000.*

FAO, 2003. *The state of food insecurity in the world 2003.* FAO, Rome. [ftp://ftp.fao.org/docrep/fao/006/j0083e/j0083e00.pdf]

FAO, IFAD and WFP, 2002. *Reducing poverty and hunger: the critical role of financing for food, agriculture and rural development: paper prepared for the International Conference on Financing for Development, Monterrey, Mexico, 18-22 March, 2002.* FAO, Rome. [ftp://ftp.fao.org/docrep/fao/003/y6265e/y6265e.pdf]

Hall, R. and Jones, C., 1999. Why do some countries produce so much more output than others? *Quarterly Journal of Economics,* 114 (1), 83-116.

Hertel, T.W., Preckel, P., Cranfield, J., et al., 2004. *The earnings effects of multilateral trade liberalization: implications for poverty in developing countries.* GTAP. GTAP Working Paper no. 16. [https://www.gtap.agecon.purdue.edu/resources/download/2874.pdf]

Hertel, T.W. and Winters, L.A., 2005. *Putting development back into the Doha agenda: poverty impacts of a WTO agreement.* World Bank, Washington.

Hoekman, B., Ng, F. and Olarreaga, M., 2002. *Reducing agricultural tariffs versus domestic support: what's more important for developing countries? Presented at the International Agricultural Trade Research Consortium conference on "The developing countries, agricultural trade and the WTO", Whistler, June 16-17, 2002.*

Kohl, R. (ed.) 2003. *Development Centre Seminars: globalization, poverty and inequality.* OECD, Paris.

OECD, 2003. *Agricultural trade and poverty: making policy analysis count.* OECD, Paris.

Pingali, P. and Stringer, R., 2003. *Food security and agriculture in the low income food deficit countries ten years after the Uruguay Round: invited paper presented at the International Conference Agricultural policy reform and the WTO: where are we heading? Capri (Italy), June 23-26, 2003.*

Stern, N., 2002. *Making trade work for poor people: speech delivered at the National Council of Applied Economic Research, New Delhi, November 28, 2002.* [http://lnweb18.worldbank.org/SAR/sa.nsf/attachments/nst/$File/j8.pdf]

Tokarick, S., 2003. *Measuring the impact of distortions in agricultural trade in partial and general equilibrium.* IMF, Washington. IMF Working Paper no. 03/110. [http://www.imf.org/external/pubs/ft/wp/2003/wp03110.pdf]

Watkins, K., 2003. *Northern agricultural policies and world poverty: will the Doha Development Round make a difference? Paper presented at the Annual Bank Conference of Development Economics, 15-16 May 2003.* Oxfam, Oxford. [http://wbln0018.worldbank.org/eurvp/web.nsf/Pages/Paper+by+Watkins/$File/WATKINS.PDF]

Winters, L.A., 2002. *Trade and poverty: is there a connection?* WTO, Geneva. [http://www.wto.org/English/news_e/pres00_e/pov3_e.pdf]

World Bank, 2005. *World development indicators 2005.* World Bank, Washington. [http://devdata.worldbank.org/wdi2005/Cover.htm]

WTO, 2003. *World trade report 2003.* WTO, Geneva. [http://www.wto.org/English/res_e/booksp_e/anrep_e/world_trade_report_2003_e.pdf]

CHAPTER 4

PRICE INTERVENTION IN SUB-SAHARAN AFRICAN AGRICULTURE

Can an institutionalist view alter our conception of the costs and benefits?

ANDREW DORWARD, JONATHAN KYDD
AND COLIN POULTON

Centre for Development and Poverty Reduction, Centre for Environmental Policy, Imperial College, London, UK

INTRODUCTION

The paper discusses the current status of African smallholder agriculture and agricultural policies. The fundamental transactions challenges in agricultural development are identified and analysed and policies which might be able to overcome these are discussed. The general approach is to build on the orthodox (or neoclassical) explanations which focus on market structures, prices and technologies, thereby bringing into view other important factors, which are the costs which agents (e.g. farmers and those who trade with or finance them) face in trying to reduce transactions risks. The risks considered are those of coordination failure, opportunism and rent seeking. It is argued that when these issues are analysed jointly with the matters explored in orthodox analysis, then successes and failures in smallholder development can be better understood. Failures may be due to the existence of 'low level equilibrium traps', which, perhaps, could be overcome by early-stage government action, but presently are not addressed by government, either because this runs counter to the prevailing ideological and policy consensus and/or due to very weak government capacity.

The picture which emerges is one in which successful smallholder development is neither a 'miracle of the market' nor predominantly the responsibility of the state. Rather, early-stage state intervention tends to be needed, not just to supply the well recognized public goods of communications infrastructure, property rights, technology research, and market and technical information. In addition,

N. Koning and P. Pinstrup-Andersen (eds.), Agricultural Trade Liberalization and the Least Developed Countries, 51–66.

governments need to engage in activities described below as 'extensive coordination', 'pump-priming' and 'threshold shifting'. This is not because states in poor countries are particularly competent in these activities (they are not!) but because often in the absence of state intervention no other actor will have the combination of scope, resources and incentives to address these critical sources of transactions failure. Ironically as agricultural development occurs, the need for government intervention is reduced, as markets begin to work better without intervention, i.e., 'thicken'. In developed economies, the emerging roles of government in agriculture tend to be concerned with the protection of consumers, the environment and landscape value. The elements of government intervention in OECD-country agriculture which are controversial, i.e. policies which support producers, can be seen as currently dysfunctional, but perhaps at earlier stages these were appropriate instruments to enable agriculture to develop. In summary, in the process of agricultural development, direct government intervention to overcome market failure is needed most in the early stages when conditions for its supply are least favourable. The policy implication is that we need to understand clearly what is needed from government at each stage of development and then focus research on the implications for politics and governance.

Our arguments concerning the fundamental requirements for early-state agricultural development, an achievement essential for the reduction of extreme poverty, have implications for debates about price intervention in poor country agriculture. Where low-level equilibrium traps are widespread due to very limited investments in the supply chain, then incentives to invest must be increased, and key means to this end are the reduction of the various sources of risk. Sometimes private agents are able to reduce risk without government, although through arrangements which are far from being endorsed by the dominant policy paradigm. The unconventional aspect is that risk is often reduced by interlocking, normally of output market contracts with those for inputs and finance, this being achieved by *de facto* monopolies and/or highly unequal power relations. Post liberalization, these arrangements have emerged for certain cash crops, notably smallholder cotton. But there are a number of critical crops for which private interlocking arrangements are difficult to achieve, so forms of 'state interlocking' may be required. It appears that forms of state interlocking (some involving private intermediaries who have direct contact with farmers) have been central to the earlier stages of most smallholder-based agricultural revolutions (e.g., the Chinese and Indian Green Revolutions).

Risks inhibiting investment in supply chains are not only those inherent in market price volatility but also the multiple potential causes of transactions failure. These may call for intervention in certain output markets, e.g. floor prices to limit risks to smallholder producers of marketed surpluses of staple food crops. Furthermore, price intervention is probably unavoidable where state interlocking marketing arrangements are needed.

AGRICULTURAL IN SUB-SAHARAN AFRICA: CHALLENGES OF GEOGRAPHY, GLOBAL MARKETS AND POLITICS

Smallholder agricultural development plays a central role in poverty reduction in Sub-Saharan Africa (SSA) due to the large numbers of poor people in rural areas and the critical role of agriculture in driving growth in poor rural areas. However, the performance of SSA's agricultural sector over the last 30 years has been disappointing. Low rates of growth in the 1970s were followed by increases in the 1980s and 1990s, but per-capita growth has been very low or negative over much of the period and SSA is the only region with agriculture growing at a rate below overall population growth from 1965-1998. This is associated with high incidence and severity of rural poverty, widespread reports of agricultural stagnation, and low use of fertilizers and low crop yields. An issue of particular concern is the reliance of much of SSA's agricultural growth on expansion of cropped areas rather than of crop yields, particularly in cereal production. SSA's increased cereal area is accompanied by a fall in rates of fertilizer use and only a slight rise in cereal yields. This pattern of agricultural change presents a major problem as cultivation extends onto increasingly fragile and vulnerable land (see for example Kydd et al. 2004).

It is widely recognized that SSA agriculture needs a process of 'sustainable intensification' with increased marketed production from greater use of purchased inputs (especially seeds and inorganic fertilizers), often with complementary use of locally available organic inputs (see for example Reardon et al. 1999). Such a pattern of sustainable intensification would provide a sounder basis for future agricultural development, but demands a framework of more complex and effective public and private institutions and faces many challenges. These challenges may be considered under three main categories: (a) arising as a result of agro-ecological and geographical conditions; (b) arising from global economic conditions; and (c) arising from political and historical conditions. Although some technical, social and political developments do offer new opportunities for growth and development, conditions for agricultural development in SSA today are generally more difficult than those that were faced by countries (mainly in Asia) which successfully developed smallholder agriculture in the past.

High on the list of SSA's challenges in agricultural development are geography and agro-ecological conditions. Most of SSA lies within the tropics where soils are often more fragile and less fertile than in temperate zones, whilst pests and diseases are harder to control. SSA also has a very varied agro-ecology with different conditions often existing side-by-side in the same country and demanding different policies, services and technology development. Thus it is difficult to generalize across and even within countries, and policy analysis and recommendations have to be tailored to match differing conditions. Variable rainfall and drought are problematic in many parts of SSA and a frequent cause of crop failure. Water control is also difficult and irrigation very limited. Partly as a result of this there are large parts of SSA where the dominant staples are roots, tubers, bananas/plantains and lower-yielding cereals such as sorghum and millet. Sustainable intensification of these crops, and of extensive livestock keeping which dominates some areas, faces a range of technical, marketing and economic challenges which are less acute with the 'green revolution cereals' (wheat, rice and maize) on which successful agricultural development was based in Asia (although maize is of course a major crop in SSA, and rice is also important in some areas).

Another major set of policy challenges arises from global economic conditions as compared with those facing countries that successfully achieved significant smallholder agricultural development in the second half of the 20th century. World export crop and food prices have fallen in real terms over the last 30 years with more integrated global markets making import substituting agriculture in SSA increasingly vulnerable to international competition. Furthermore, global markets reward supply chains characterized by rapid information exchange, flexible response, quality control and tracking. This tends to undermine the historic competitive advantages of labour intensive agriculture small farming. Smallholders are perhaps least disadvantaged by these developments in supplying their national markets with tradable and semi-tradable staples and vegetables and fruits. They are also low-cost producers of certain traditional commodity exports (e.g. cotton, cocoa and sugar) but generally face tough conditions in markets which are often distorted by rich-country agricultural support.

Many parts of SSA face particularly difficult policy and political constraints. Formal political structures and institutions tend to be relatively new, with substantial political change since independence. Countries tend to be culturally and ethnically diverse. Patrimonial systems of government and politics are common, diverting resources from broader development goals without effective checks and balances, and without a sizeable and well established middle class providing a strong administrative cadre holding governments to account. Smallholder farmers also tend to be a weakly organized and represented constituency, despite their large numbers.

POLICY CHALLENGES: AGRICULTURAL INTENSIFICATION AND COORDINATION IN SUPPLY CHAIN DEVELOPMENT

The last thirty years have seen dramatic changes in the dominant economic policy paradigms among international organizations and OECD countries, with increasing scepticism as regards the effectiveness of state agencies as economic actors and increasing emphasis on market solutions. A consequence of this reverse in the ideological climate is that pre-liberalization policies which supported some very successful agricultural development in other parts of the world (particularly Asia) in the latter parts of the 20th century have not been available to SSA governments over the last 20 years or so as a result of both their own inability to pursue them without external financial support, and the promotion by international financial agencies of liberalization policies.

Agricultural intensification involves technical change and marketing systems to supply inputs and seasonal finance. Intensification therefore involves the development of supply chains around smallholder farmers[1], with simultaneous and complementary investments in all links in the supply chain. Making these simultaneous investments can, however, pose serious difficulties in poor rural areas, as a result of transaction costs and risks, which include coordination, opportunism and rent-seeking costs and risks.

Coordination problems in poor rural economies

Poor rural areas within low-income economies are characterized by low total and monetary incomes for most people, with consequent limited consumption and expenditures, a weakly developed monetary economy with a narrow base, and markets (for agricultural inputs, outputs and finance, consumer goods and services, etc.) which are relatively 'thin' (with small volumes traded, although for some items there may be very large numbers of people trading in very small volumes) and prone to large seasonal variability in demand and supply. These conditions normally coexist with poor roads and telecommunications; poor information (particularly in agriculture, on prices, on new technologies, and on potential contracting partners); difficulties in enforcing impersonal contracts; and rent-seeking behaviour by politicians, bureaucrats, criminals and the private sector.

These conditions pose particular problems for the supply chain development needed for agricultural intensification, as this requires significant investments by new players entering the market, investments which carry high risks of transaction failure and (the other side of the coin) high transaction costs involved in obtaining protection against such risk. These risks and costs can be considered in terms of rent-seeking, coordination and opportunism.

- Coordination risks are the risk of an investment failing as a result of the absence of complementary investments by other players in a supply chain.
- Opportunism risks arise when another contracting party, with monopsonistic or monopolistic control over a complementary investment or service, removes or threatens to remove it from the supply chain after a player has made an investment that depends upon its continuing supply.
- Rent-seeking risks arise when powerful government, political, criminal or other agents not directly party to a transaction see associated investments and/or revenue as an opportunity to expropriate or threaten to expropriate income or assets from the investor.

Coordination, opportunism and rent risks (and the costs of protection against them) are closely related, and when these are high as compared with potential returns to investment, then the potential investors required to establish new activities for developing an agricultural intensification supply chain may find the investments too risky, and thus the supply chain may not develop, even if it is potentially profitable (once necessary investments have been undertaken).

This situation is described in a formal economic model in Figures 1 and 2, which describe a situation where all actors face a two-stage investment problem: they must make stage-1 investments in assets specific to a particular supply chain activity in order to reap net revenues in stage 2. Their revenues in stage 2, however, are determined not only by the scale of their own stage-1 investments, but also by the scale of others' stage-1 investments (investments which are not known to them when they make stage-1 investments).

Figure 1 shows the relationship between individual actors' marginal factor costs and marginal value products (on the vertical axis) from seasonal investments, under conditions of different behaviour by other actors in the supply chain, taking smallholder maize production in a poor rural area as an example. This diagram shows that investment in seasonal inputs (stage-1 investments) without complementary investments and transactions (by input sellers, financiers and produce buyers) incurs high marginal factor costs (MFC_0) and a rapidly falling marginal value product (MVP_0). The result is profit maximization around subsistence production (with investment I_0), and only small surplus sales in good and normal years. With complementary investments and transactions by other actors, however, reduced transaction costs and risks lead to a fall in marginal factor costs to MFC_1, and the marginal value product is maintained for surplus sales and hence higher production (MVP_1). The combination of lower MFC and higher MVP leads to profit maximization at much higher levels of investment (I_1) and net income, with a significant marketable surplus beyond the households' own subsistence needs.

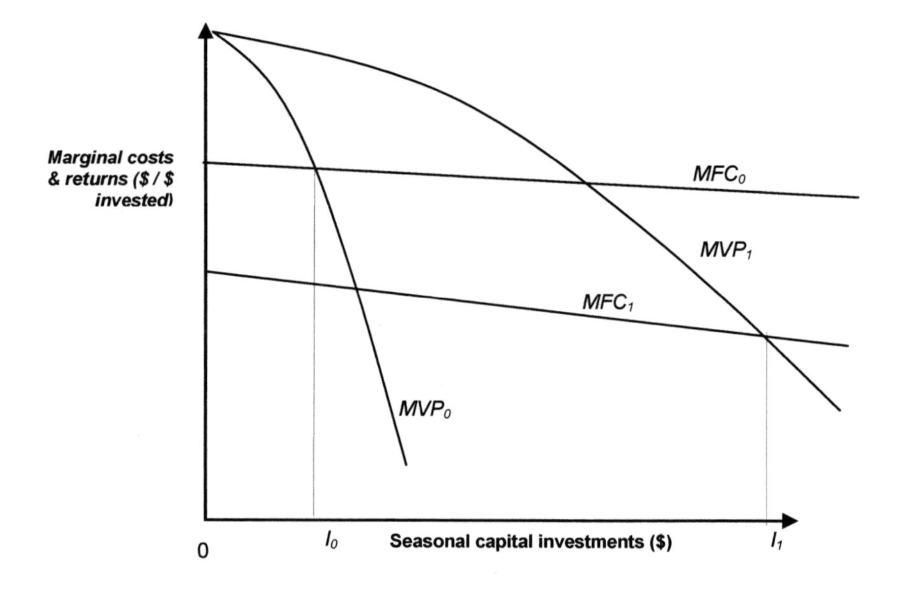

***Figure 1**. High and low level firm investment equilibria*

If a similar situation is faced by the other actors making complementary investments in the supply chain, then there will be two possible system equilibria as shown in Figure 2. This examines marginal factor costs and marginal value products for investment in an industry or commodity supply chain assuming that this is distributed along a complete supply chain. It distinguishes between different elements of marginal factor cost (MFC). We begin by considering only conventional neo-classical production economics analysis, using a 'Base MFC' line, which is determined by factor prices[2]. Considering only these factor prices, optimal supply chain investment occurs where the Marginal Value Product (MVP) curve cuts the Base MFC line, at E. The shape and position of the MVP curve is determined by the price of the supply chain output(s) and by the technologies employed (higher prices and better technologies both lift the MVP curve, while diminishing marginal returns and falling prices in limited markets both cause MVP to fall at higher levels of investment).

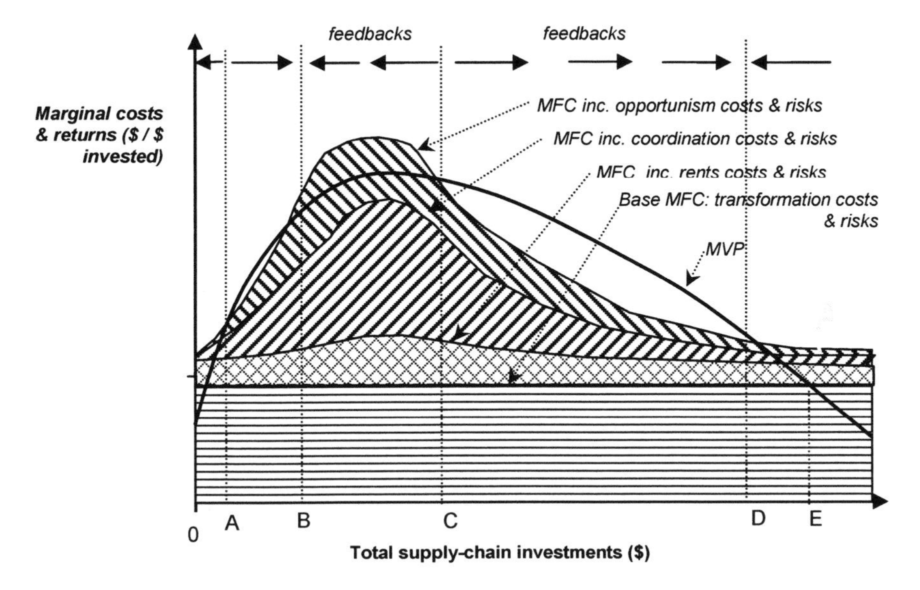

Figure 2*. High and low level supply chain equilibria*

We now introduce costs and risks associated with coordination failure, opportunism and rent seeking. These are represented in Figure 2 in three bands above transformation costs and risks.

The second cost and risk band in Figure 2 represents rents. There is a long standing and increasing concern about poor governance and opportunities for elites (for example politicians, civil servants or formal or informal groups or individuals) to extract 'rents' in the context of weak or poor and predatory governance systems. These rents may be legitimate tax demands or illegitimate demands for bribes, 'cuts' or 'fines'. Rents can have positive effects (for example financing delivery of public goods and/or accumulation of capital for local investment or redistribution as described by Khan 2004) but these positive effects (where they exist, and in many cases they do not) need to be set against their costs: increased risks, uncertainty and costs in productive activity, with depressed and distorted returns to and incentives for investment. There are no strong *a priori* arguments for a particular relation between total supply chain investment and MFC for rent costs and risks, but one might expect MFC to decline with increasing supply chain investment (*ceteris paribus*).

The third and fourth cost and risk bands in Figure 2 represent coordination and opportunisms risks and costs. The nature of the relation between thin markets on the one hand and risks and exposure to coordination failure and opportunism on the other suggests that large levels of investment in a supply chain should substantially reduce coordination and opportunism costs and risks. Reduced risks of coordination failure and opportunism (and hence falling MFCs) are likely at high levels of investment either through thick markets (as discussed earlier) and/or through efficiencies achieved in large firms (an issue we discuss later)[3]. Reduced risks mean that less costly counter measures are required, but unit transaction costs also fall with higher volumes, giving a double benefit in cost reduction from greater levels of investment and turnover[4].

The most obvious impact of adding coordination, opportunism and rent costs and risks to the conventional neo-classical analysis is a shift of the profit-maximizing equilibrium point to the left (from point E to point D), leading to lower levels of investment and production. There is also a very substantial shrinkage of the region where MVP is greater than MFC (between investment levels C and D). If investments in a supply chain are initially below C, then investors have no immediate gains from increased investment (since MFC is greater than MVP) and no incentives to invest – in fact the incentive is to reduce investment as long as MFC is greater than MVP. As drawn, this will cause investment to fall to B, which represents a low-level equilibrium (equivalent to profit maximization around subsistence production)[5]. There is then a critical threshold level of total supply chain investment (point C in Figure 2) below which the marginal returns to investment are negative. The total level of investment therefore has positive (or negative) feedbacks above (or below) this threshold. Below the threshold the supply chain is caught in a low-level equilibrium trap.

This analysis depends upon two conditions: (a) individual players facing different individual MVP and MFC curves depending upon total (balanced) supply chain investment (as shown in Figure 1)[6]; and (b) some institutional coordination failure that prevents players individually or collectively moving to high levels of supply-chain investment. Generally, smallholder farming areas of SSA are characterized by an atomistic market, with many small players but without non-market coordination or significant efforts towards collective action. This analysis explains individual choices around a stable low-level equilibrium: ironically (given the debates about market liberalization) the neo-classical ideal of perfectly competitive markets then provides some of the necessary conditions for coordination failure, and escape from the low-level equilibrium trap requires the development of non-market coordination mechanisms.

Williamson (1994; 1985; 1991) identifies firms, markets and relational contracts (or hierarchy, market and hybrid arrangements) as the three main types of contractual arrangement, with widespread use of hierarchy and hybrid arrangements to deal with problems of asset specificity in developed economies. Hall and Soskice (2001), comparing the relative importance of hybrid and competitive market arrangements in different OECD economies, highlight first the importance of large firms and hierarchical arrangements in providing coordination mechanisms in all types of market economy and second the comparative institutional advantages of greater reliance on non-market arrangements for coordination between firms in industries where large investments are needed in specific assets[7]. Both these points challenge simplistic prescriptions for the development of markets as a necessary component of efficient economic development.

The observation that large firms and hierarchical arrangements play a major role in all types of developed market economy contrasts with the lack of large firms and hierarchies in many poor economies (Fafchamps (2004) demonstrates this very clearly for SSA economies). It also suggests that the increased coordination required for economic growth and development tends to be delivered by a shift from poorer economies dominated by small atomistic players linked by (weak) market and hybrid arrangements to greater reliance on thicker markets and/or hybrid arrangements linking larger firms in wealthier economies. More developed economies are therefore characterized by increased scale and scope of hierarchical arrangements. This represents an important challenge to neo-classical orthodoxy, as it suggests that the development of larger hierarchy arrangements may be at least as important in economic growth as the development of wider competitive market arrangements. Development should then be characterized not in terms of development of a market economy but as a movement from '*atomistic relational market systems*' to '*market and hierarchy reputational systems*'.

Why do wider hierarchical arrangements often fail to develop to overcome the associated asset specificity and low-level equilibrium trap problems of poor rural areas? Hybrid arrangements are common in poor rural economies, but usually involve relational contracts between individuals or small firms (Fafchamps 2004) and thus tend to be limited in the scale and geographical scope of their activities. A number of factors inhibit both endogenous development of larger firms and inward investment by large urban-based or foreign-owned firms: difficulties in acquiring large areas of land in poor rural areas; particular difficulties in coordination without control over agricultural land and production; a large minimum scale needed to achieve the levels of supply chain investment and activity required to cross the low-level equilibrium threshold (preventing the growth or endogenous development of firms); poor communications infrastructure; weak institutional environment and property rights; limited numbers of people with entrepreneurial skills and local and personal knowledge; costly and difficult access to capital; and high risks and relatively low returns compared to alternative investment opportunities. The last point is particularly applicable to food crops[8]. As a result, although there have been many large-scale inward cash crop investments by large firms, there are very few private investment success stories in smallholder food crop production without substantial public sector support[9]. This observation is of substantial importance as food crops constitute a major and critical part of poor rural economies, and historically their development has provided the initial stimulus to most examples of successful pro-poor growth in poor rural economies.

Policies for overcoming coordination failure in poor rural economies

This analysis of the development challenges posed by thin markets, asset specificity and coordination failure has practical implications for policies promoting market-led pro-poor agricultural growth in poor rural areas as we can use it to consider processes by which a set of actors may escape from the trap (and increase productivity at higher equilibria). We use the broad structure of Figure 2 to identify three broad 'functions' of development interventions:

- supply chain coordination (allowing investment decisions to transcend the narrow self-interests of different players in the supply chain);
- pump priming investment (lifting supply chain investments across critical minimum thresholds);
- threshold shifting (which involves changing the MVP and different MFC curves to move or remove thresholds).

The first intervention 'function' involves the development of an effective system supporting coordinated, complementary decision making by different players across a supply chain. The major alternative forms of institutional arrangement which such a system may use for achieving this have already been discussed (market, hierarchy and hybrid arrangements) and it is clear that a system relying predominantly on market mechanisms will not be able to provide the coordination necessary to cross substantial thresholds – although market mechanisms may have more of a role where the thresholds themselves can be removed or substantially reduced as part of the broad transition from an 'atomistic market and relational economy' to alternative forms of 'market and hierarchy economy' discussed earlier.

Kydd and Dorward (2004) classify non-market coordination systems in terms of 'local' and 'extensive' *scope* of coordination and 'exogenous' and 'endogenous' *processes* of coordination development. Endogenous 'local' coordination systems may develop either through replacement of smallholders by larger-scale (private or state) farms or through local relations linking different local agents interested in investing in different activities in the supply chain, for example through farmer groups or through interlocking arrangements by (generally powerful) traders. In staple crops, where total supply chain profits are likely to be more limited than in cash crops, progress in local investment is likely to be slow (as low returns weaken both the incentives to set up coordinating institutions and the penalties for defection). Eventually, however, if there is sufficient growth in local coordination arrangements then these may in aggregate reach the threshold level of total investment in the supply chain, enabling a transition into a market and hierarchy based coordination system and growth path. Left to itself this process is, however, likely to be slow and fragile, highly path dependent and susceptible to political economy processes of rent seeking and to shocks affecting the total investment threshold.

Exogenous alternatives to slow and fragile endogenous local coordination processes are (a) externally assisted 'soft' local coordination processes (for example involving state or NGO support for the development of farmer organizations, for trader associations, or for contract grower, nucleus/ outgrower and other interlocking systems); or (b) more extensive 'hard' coordination where a strong central coordinating body with a mandate from the state ensures investments across the supply chain with highly credible coordinated commitments[10]. As discussed later, agricultural parastatals in SSA often attempted to follow this last approach by establishing large hierarchical organizations (large in scale and scope). These large parastatal hierarchies then (with government agencies) took over investments and investment risks for all parts of the supply chain except on-farm production and retail sales (although even here they were sometimes involved), and then tried to establish links with farmers to constitute a major part of a coordinated system for planning and delivery of farmer services (for financial services, and input and output marketing).

The parastatal system is not the only model for pursuing 'extensive coordination' but it is a highly instructive one in many ways. Its dramatic failures and achievements highlight both the difficulties facing the development of extensive coordination and the potential for success. Furthermore, where it was successful, it generally involved not only effective action to improve supply chain coordination (the focus of our discussion above), but also action to support the two other 'escape mechanisms' discussed earlier and to which we now turn: pump priming and threshold shifting. This reflects a simple conclusion from the relationships illustrated in Figure 2, that the development of coordination mechanisms (through endogenous local mechanisms or through different types of local and extensive exogenous external support) will be easier the closer a supply chain is to its critical threshold (at C in Figure 2), and this situation will arise with a higher investment base and/or higher profits in the supply chain.

The second function for development interventions, 'pump priming investment', seeks to provide this higher investment base. It involves government or donor investments attempting to move the level and density of investment in an economy, sector or supply chain to the right and beyond or near the critical threshold at point C in Figure 2. Attention needs to be paid here to types and modes of investment and/or subsidy that are effective in promoting substantial thickening of markets and increases in economic activity. Important challenges concern (a) identifying critical elements of a supply chain where investment will have wider stimulative effects (allowing for complementarity between some of these); and (b) ensuring that pump priming is large enough and continues long enough to cause major and permanent shifts in expectations and structural relations within the supply chain while (c) investing in ways that promote complementary private sector investment rather than crowding it out or inhibiting it; and (d) also establishing strict and clear rules establishing time and fiscal limits to public sector investment. Historically the sustained green revolutions in Asia have been successful with (a), (b) and (perhaps to a lesser extent) (c) above, whereas the more abortive green revolutions in SSA have only achieved the first of these, and have then been forced to discontinue investments for reasons of ideology and/or fiscal constraints[11]. Establishing time and fiscal limits to public sector investment is almost universally problematic (as the agricultural policies of most OECD countries demonstrate), but the critical challenge for developing countries is to ensure that the costs do not rise so rapidly as to present a fiscal crisis before major and permanent shifts have been achieved in expectations and structural relations within the supply chain.

Pump priming investment will not have to achieve so much and improving coordination systems will be easier if the critical total supply-chain investment threshold (point C in Figure 2) is lower. Threshold shifting, the third broad development function identified earlier, is represented in Figure 2 by movement of the MVP curve upwards and of the MFC curves downwards so that point C moves to the left (to lower levels of investment) or disappears altogether. Even without any low-level equilibrium trap (i.e. in the absence of point C) upward MVP shifts or downward MFC shifts are beneficial as they will lead to increased supply-chain profitability and higher equilibrium investment with higher production.

An upward shift of the MVP curve may be achieved by technical change (with increases in marginal productivity of investment) or by increases in output price. This represents the focus of part of current policy orthodoxy's emphasis on technical change from agricultural research and extension and better producer prices from structural adjustment. Technological development, however, generally requires coordination between different links in increasingly complex supply chains, with increasing investment by different and growing numbers of players. Complementary action is therefore often needed to simultaneously improve coordination and promote technical change, and this needs to be taken into account in the development and promotion of new technologies.

Downward movement of the MFC curves may be achieved by reduced input prices and costs (reducing transformation costs) or by reducing costs and risks of coordination failure, opportunism or rents. Again current policy orthodoxy emphasizing technical change from agricultural research and extension looks to reduce transformation costs and risks in the base MFC while more recent policy emphasis on promoting institutional and property rights development seeks to reduce the costs and risks of opportunism and rents and implicitly looks to the development of competitive markets to reduce coordination costs and risks (although our arguments suggests that under certain circumstances this reliance on competitive markets to reduce coordination costs and risks may be misplaced).

It is important to note here a useful if not always clear distinction between improvements in overall supply chain coordination (which were discussed earlier and are concerned with development of broader coordinating systems) and specific cost reducing institutional arrangements between different players within such a system. Both are needed, the latter being important for actually delivering different systems' potential reductions in coordination and opportunism costs and risks. Here the detailed structuring of relations is important (for example interlocking of bilateral transactions, or the organizational structures and staff management and incentive systems within hierarchies), echoing an important point made by Omamo (2003) that modalities of how policies are implemented are often more important than the finer points of what policies to implement.

Implications for trade policy

Our arguments have strong implications for international and domestic trade policies. The three categories of development interventions (supply chain coordination; pump priming investment; threshold shifting) imply government intervention.

Measures to promote supply chain coordination may require government agencies (or private agencies contracted to and/or regulated by government) to intervene in markets, sometimes restricting the freedom of agents in the chain. The most obvious example is the requirement to maintain interlocking transactions where monophony may be enforced to counter opportunism and consequent supply chain failure. Pump priming investment may require government funds to provide infrastructure, subsidize services and invest in processing in the expectation of low financial rates of return. Threshold shifting implies price interventions (for inputs and output) which alter the value and cost functions of Figure 2 in order to bring particular areas of smallholder agriculture to the 'right side' of 'BC', the 'zone of market failure' illustrated in the diagram. To the 'right side' of C there is the prospect of dynamic development: as further increases in volume will be profitable for agents in the system and so further expansion will occur. This may stimulate economies of scale and it is possible that the cost function will diminish with scale to an extent much greater than is suggested by Figure 2.

Price interventions could take the form of subsidies to producers, but for administrative and budgetary reasons will probably require a mix of border measures and subsidies. Border measures (tariffs etc) would be required to shift up the MVP of importable, and also to stem a possible flow of subsidized inputs.

A key question is whether SSA countries are prevented by WTO rules and/or donor conditionality from pursuing these policies. Lockwood (2005, pp. 39-44) argues that WTO rules are not as constraining as many NGO campaigns tend to imply. The fault may more with (i) donor ideology and conditionality; (ii) SSA governments' understanding of what has to be done; and (iii) their political will and competence.

WHERE FROM HERE?

This paper has explained why in specified but common circumstances state intervention is needed to enable smallholder development which goes beyond the supply of public goods, conventionally defined. Often, for the most critical crops from the point of view of poverty reduction, semi-tradable staples, government intervention is needed to provide coordination and otherwise reduce risks faced by investors in agricultural supply chains (a category which includes farmers). For example, government may have to provide a framework for, and financially guarantee, state interlocking as a form of 'extensive coordination'. Within the types of arrangement, price floor and possibly price maxima are likely to be unavoidable features.

The paper has not sought to describe in any detail the forms which price intervention might take. Clearly there are huge and well-known pitfalls to be negotiated, including: unacceptable fiscal costs; effects on poor consumers; inefficient resource allocation; rent seeking; and the generally weaker performance incentives which exist in government. In Sub-Saharan Africa borders are *de facto* very open, and any government offering to buy staples at well above the regional price will could end up stockpiling its neighbours' product. Our view is that the challenge is to find models which control and try to minimize these highly undesirable consequences of intervention, rather than to regard them as sufficient reasons in themselves for there to be no forms of price intervention. Our preliminary thinking on these matters has taken us in both micro and macro directions. On the micro-side, is it possible to design interventions which will be relevant to defined areas of production, e.g., where there is a high concentration of smallholders who would benefit? On the macro-side, should African government join together to create regional 'common agricultural policies', for example for Western, Eastern or Southern Africa? Regional CAPs could limit the porous border arguments against intervention, incorporate areas with non-covariant production risks and perhaps somewhat depoliticize (in term of national politics) hard decisions concerned with food security, consumer welfare and farmer incomes.

NOTES

[1] There is a large literature about the importance of smallholder agriculture in driving pro-poor growth; see for example Kydd et al. (2004) for a recent discussion.

[2] Where there are returns to scale in purchasing or transport costs then the MFC may be slightly downward-sloping, but otherwise in perfectly competitive markets; the Base MFC should be roughly constant and independent of scale. A supply chain may, however, constitute a substantial share of input markets, and in such circumstances the Base MFC would be expected to rise with increasing supply-chain investments. The slope and shape of the Base MFC are therefore likely to vary between different situations.

[3] Transaction risks in market arrangements are likely to fall at higher levels of supply-chain investment as more players allow market coordination mechanisms to work and reduce the risks and costs of protection against both coordination failure and opportunism. Larger transaction volumes and/or more frequent transactions also reduce costs and risks in (inherently less risky) hybrid and hierarchical arrangements for exchange and coordination as the fixed costs of establishing these relationships are spread over larger and more frequent transactions, and more frequent transactions themselves facilitate the establishment of these relations and provide incentives for contracting parties and employees to honour them (Williamson 1985; 1991).

[4] Discussion of Figure 2 focuses on declining rent, coordination and opportunism costs at higher levels of supply-chain investment, as this is critical to understanding coordination failure and the low-level equilibrium trap. In some circumstances, however, low levels of investment may support very local or within-household production and consumption chains. In such circumstances increasing investments may face increasing MFCs from risks of coordination failure and opportunism due to the crossing of thresholds from subsistence to surplus production and sales (by individual households and by local communities), leading to the need for widening circles of trade and hence of trading relationships. In the context of a weak institutional environment and thin markets, the establishment of new trading relations carries significant costs and risks. This postulated behaviour of the MFC curve at low investments (as drawn in Figure 2) is not critical to the basic conceptualization of low-level equilibrium traps, it merely explains the existence of *non-zero* low-level equilibria. The high but falling MFC at higher levels of investment is, however, critical to the existence of low-level equilibrium traps.

[5] As noted in the previous footnote, at low levels of investment the MFC and MVP curves may take a variety of different shapes, and relate to each other in a variety of ways. The broader argument for the existence of a low-level equilibrium trap is not sensitive to these shapes provided that with increasing total supply-chain investment MFC moves from a position above MVP to one where it lies below the MVP, before these positions are again reversed. In other words, crossover points C and D are critical to the existence of high and low equilibria. Drawing of crossover points A and B in Figure 2 illustrates ways in which non-zero low-level equilibria may exist, but this is not critical to the coordination-failure arguments developed in this paper.

[6] The differences in Figure 2.1 between MFC and MVP in the presence and absence of assured complementary investments and transactions result from differences in these costs and risks in input and finance markets (for the MFC curves) and in output markets (for the MVP curves) There may also be differences in technology, where a low-input technology is more profitable under high-risk/cost conditions and a high-input technology is more profitable under low-risk/cost conditions. This is particularly relevant for sustainable intensification in smallholder agriculture.

[7] Hall and Soskice (2001) distinguish between liberalized market economies (LME's) and coordinated market economies (CMEs). In the first case liberalized markets provide the main coordinating systems between firms while in the second case coordination is also achieved through significant state activism and/or through membership associations linking different firms engaged in common supply chains.

[8] Many of these problems are less severe for some cash crops needing large but potentially very profitable investments in processing facilities. These investments provide foreign companies with profit incentives to invest in interlocking systems for vertically integrated coordination of seasonal input and finance and other services needed to induce sufficient and reliable smallholder production to make the investment in processing facilities profitable. Critically, however, the need for large-scale investments also makes it easier to develop institutional arrangements protecting investments in seasonal finance delivery against opportunism by farmers and crop traders. This is because large foreign firms have greater ability to access external sources of capital and expertise needed for investments in processing facilities, and this can provide them with a monopoly over crop-processing facilities, and so control over the supply chain.

[9] Even in cash-crop production systems, some government or donor coordination or subsidy has often played a part in attracting foreign investment.

[10] This distinction between 'soft' and 'hard' promotion of coordination reflects observations by Hall and Soskice (2001)of differences between CMEs in types of state support.

[11] Even where fiscal constraints forced policy changes, the prioritization of fiscal cuts often reflected dominant donor ideologies.

REFERENCES

Fafchamps, M., 2004. *Market institutions in Sub-Saharan Africa: theory and evidence.* MIT Press, Cambridge.

Hall, P.A. and Soskice, D.W., 2001. *Varieties of capitalism: the institutional foundations of comparative advantage.* Oxford University Press, Oxford.

Khan, M.H., 2004. *Strategies for state-led social transformation: rent management, technology acquisition and long-term growth: paper presented at ADB workshop on making markets work better for the poor, Hanoi, Vietnam 12-15 April 2004.* Asian Development Bank.

Kydd, J. and Dorward, A., 2004. Implications of market and coordination failures for rural development in least developed countries. *Journal of International Development,* 16 (7), 951-970.

Kydd, J., Dorward, A., Morrison, J., et al., 2004. Agricultural development and pro-poor economic growth in sub-Saharan Africa: potential and policy. *Oxford Development Studies,* 32 (1), 37-57.

Lockwood, M., 2005. *The state they're in: an agenda for international action on poverty in Africa.* ITDG Publishing, London.

Omamo, S.W., 2003. *Policy research on African agriculture: trends, gaps, and challenges.* International Service for National Agricultural Research, The Hague.

Reardon, T., Barrett, C., Kelly, V., et al., 1999. Policy reforms and sustainable agricultural intensification in Africa. *Development Policy Review,* 17 (4), 375-395.

Williamson, J., 1994. *The political economy of policy reform.* Institute for International Economics, Washington.
Williamson, O.E., 1985. *The economic institutions of capitalism.* Free Press, New York.
Williamson, O.E., 1991. Comparative economic organization: the analysis of discrete structural alternatives. *Administrative Science Quarterly,* 36 (2), 269-296.

CHAPTER 5

POVERTY, LAND CONSERVATION AND INTERGENERATIONAL EQUITY

Will the least developed countries benefit from agricultural trade liberalization?

KIMSEYINGA SAVADOGO

Faculty of Economics and Management, University of Ouagadougou, Burkina Faso

INTRODUCTION

The current negotiations on agriculture within the World Trade Organization (WTO) may lead to some degrees of liberalization in world agriculture. Many of the developing countries, and chiefly among them the poorest, rely heavily on agriculture for their livelihoods. A question is whether the change in the rules of trade will affect these poor countries in a positive way.

For neoclassical trade theory, free trade has unambiguous beneficial effects for all trading partners through the working of static comparative advantage. This prediction is intended to apply to all commodities, including agricultural, and to all countries, highly or less developed. If successful, sustained WTO negotiations on the liberalization of agricultural trade will produce a free trade situation where farmers compete against farmers in a global setting, replacing the old situation where countries competed against other countries through the use of agricultural trade policies and where farmers were shielded from the direct competition from other farmers. Such a new setting, according to neoclassical argument, will be beneficial to farmers in the least developed countries (LDCs) essentially through the more favourable prices of exported products that will result.

Two objections against this conclusion have been raised in the literature: (1) the presence of second round effects, such as the changing farm wages and/or production basket mix resulting from the changing prices, and (2) the existence of structural factors that may limit farmers' responses to positive price changes, such as

N. Koning and P. Pinstrup-Andersen (eds.), Agricultural Trade Liberalization and the Least Developed Countries, 67–81.

weak market integration, and credit constraints (Narayanan and Gulati 2002). A third, and less frequently discussed, complication is the intergenerational tragedy of the commons in natural resource management that is caused by extreme poverty. In dealing with land conservation issues, land is normally treated in the literature as a pure private good, thus precluding any justification for public intervention. However, we will argue that in a context of extreme poverty, investing in natural resources such as land faces a problem of free riding when looked at across generations. That is, in the context of budget restraints, successive generations may tend to use whatever fertility is left on the land, without investing in land regeneration. If pervasive, this problem of inadequate level of private investment in land will imply that farmers may not be prepared to take full advantage of trade liberalization. We propose policies to tackle this eventuality, including a role for the international community.

In section *"Extent of land degradation"* below, we indicate the extent of land degradation in developing countries, especially in Sub-Saharan Africa. In section *"Land degradation and sub-optimal private response in an intergenerational context"*, we develop a conceptual argument that explains the inadequacy of private choices in land fertility restoration. In section *"Land regeneration in the LDCs: the role of prices and incomes"*, the links between land degradation, prices, income and public policy are discussed and illustrated with factual evidence from Burkina Faso. Section *"Implications for price and trade policies"* considers the role of active price and trade policies in securing benefits from liberalization for the LDCs. Section *"Summary and conclusion"* offers some conclusions.

EXTENT OF LAND DEGRADATION

Land degradation has different definitions, but one that is sufficiently comprehensive says that it is "the aggregate diminution of the productive potential of the land, including its major uses (rainfed, irrigated, rangeland, forest), its farming systems (e.g. smallholder subsistence) and its value as economic resource" (Stocking and Murnaghan 2001). Gretton and Salma (1997) add that degradation has to be caused by human activity. Desertification due to natural climate changes would not be regarded as degradation while desert-like conditions due to overgrazing would.

For soil scientists, land degradation has a significant aspect of irreversibility (Eswaran et al. 2001). Lost soil cannot be fully recovered. Besides, non-linearities in soil–crop interactions often lead to dual equilibrium situations. For example, land degradation may cause high leaching and bad rooting, which lowers the nutrient recovery rate of plants. As a consequence, the application of fertilizer has less effect, which reduces the returns on investment aimed at regenerating the soil.

An estimated 1.9 billion hectares of land worldwide are affected by degradation (El-Beltagy 1997). Dregne and Chou (1992) estimated that 70% of the land in dry areas in the world was degraded in the 1990s. In Africa, this proportion was 73%. The rate of degradation is estimated at 21 million hectares per year, with 6 million hectares of land permanently lost for crop production. The problem of land degradation is an important global issue because of its adverse impact on agricultural productivity and the environment, as well as its effect on food. Projections suggest that if the negative trends were to continue, the future food security of poor countries is threatened (El-Beltagy 1997; Eswaran et al. 2001).

Sub-Saharan Africa is a geologically old region with many poor and fragile soils. Nutrient depletion is a major form of land degradation and has severe economic impact. In a sample of 38 African countries, including 26 LDCs, average annual per hectare losses of nutrients were estimated at 22 kg of nitrogen, 3 kg of phosphorus and 15 kg of potassium (Holden 1997).

The impact of soil degradation on agricultural productivity has also been assessed. The productivity of some lands in Africa has declined by 50% due to soil erosion and desertification (Dregne 1990). Lal (1995) estimated that yield losses due to erosion in Africa ranged from 2 to 40%, with a mean loss of 8.2%. Projections show that yield reductions might be as high as 16.5% by 2020, suggesting that the land could practically lose all its productive potential if soil degradation continues unabated. Admittedly, it is difficult to obtain a non-ambiguous cause-and-effect relationship between land degradation and productivity at the micro (plot) level. Data from China and Thailand failed to provide any evidence of a negative relationship between cumulative soil loss and yield per hectare. More insight is needed into the precise nature of processes involved at the soil–plant–atmosphere continuum (Eswaran et al. 2001). Nevertheless, there is increasing evidence that the economic effects of land degradation are serious. Poor people mainly rely on agriculture, and negative changes have detrimental effects on them in the form of food insecurity, malnutrition and child mortality.

LAND DEGRADATION AND SUB-OPTIMAL PRIVATE RESPONSE IN AN INTERGENERATIONAL CONTEXT

The simple neoclassical model predicts that free market forces will lead to a Pareto-optimal situation. However, this conclusion presupposes the absence of external effects and public-good problems. One justification for public intervention in environmental protection is externalities *sensu stricto*. For example, land erosion from untended fields may lead to the silting of rivers downstream, the contamination of drinking water by agrochemicals, and loss of habitat (Scherr and Yadav 1997), causing negative externalities for all users. In this paper we follow an alternative path by stressing intertemporal externalities between individuals who use the same piece of land successively without intervention of a market. Our argument is that in the presence of extensive poverty, land that is passed on to family members of a new generation may become a common pool resource in an intergenerational sense. This is so, even when the land itself is a private (or semi-private, collectively inherited) good, provided that the land is bequeathed with little role being played by the market, as commonly occurs in African LDCs.

A common pool resource is defined as a depletable good from the consumption of which other individuals cannot be excluded. The fish in the sea is an example. The non-excludability is normally seen as space-related, not time-related; the definition stresses the simultaneous use of the good. We extend this definition to the case where investment in land has a time aspect, through the intergenerational linkages binding individuals exploiting pieces of land bequeathed through inheritance. Non-excludability in this extension means that an individual at time t cannot exclude other individuals at $t-1$ or $t+1$ from using the resource.

Poverty is the major factor that may cause an intergenerational tragedy of the commons. The standard 'tragedy of the commons' model presupposes impersonal relationships between the individuals that are using the common resource. However, the individuals who are successively exploiting a piece of family land are connected through kinship relations. The utility function of a farmer normally includes the utility of his sons or other heirs. Their future well-being is a consideration in his investment decisions, so that a tragedy of the commons will not normally arise. However, if poverty causes high individual discount rates, farmers will not only heavily discount their own future well-being, but that of their heirs as well. Moreover, insofar as care for relatives has the nature of a luxury good, poverty will reduce the weight of the heirs' well-being in a farmer's budget-constrained utility function. As a consequence, heirs of poor farmers may find themselves being treated like strangers. Their parents may choose to free ride by exploiting the land that is the common resource of successive generations without replacing the amount of nutrients that they deplete. Such intergenerational free riding is rendered possible by the fact that the productive potential of land can be 'stretched' to some extent, unlike other productive assets such as nitrogen fertilizer, which are depleted in one season.

In this way, intergenerational equity issues may become a serious concern. A Sahelian farmer now in his forties could reflect on the discourse on land degradation and rehabilitation during the Sahel drought spell of 1968-74. He would recall that it was said then, as he was a little boy, that in the next thirty years or so, failing to invest in land improvement techniques would be tantamount to suicide for the rural population, or that the countries would become dependent on the rest of the world for their food. Sitting today in his degraded gravel-laden field, such a farmer would say to himself: "I am living with this terrible natural resource deficit now because my parents failed to take my welfare into account when planning on their own needs, as I was growing up". This intergenerational memory looping, through which it is possible to look back in time, uncovers the reality of an intergenerational tragedy of the commons on privately bequeathed land[1].

One might object that because the well-being of the son is an element of the utility function of the father, and the discounting out of it follows from the latter maximizing his utility, the exhaustion of the land – no matter how inequitable for the son – is not an external effect and therefore not inefficient. However, there is no sense in which a change can be said to be efficiency improving if the losers are not compensated. That the hypothetical compensation criterion would be enough to establish an actual efficiency gain is a misunderstanding (see, e.g., Jongeneel and Koning 1999).

Poverty trap

An intergenerational tragedy of the commons implies that investment to restore land by individual farmers will be below the socially optimal level, i.e. one that would take into account the welfare of future generations. The ensuing problems can become intractable because of the poverty trap. Various conditions, not uncommon in the less developed countries, can give rise to this. One, already referred to above, is the high time preference typical of poverty, which makes poor people deplete their land. Another is the indivisibility of land conservation investment, which typically comes in packages. Anything less than the minimum package will have little effect, but the poor may be unable to afford this package. Still another condition is high production risk (weather, pests) that limits poor people's ventures into successful land conservation technologies such as the use of costly commercial fertilizers. Finally, there is the dual equilibrium nature of many soil–crop systems that makes it costly to return to the higher-level equilibrium once a low-level equilibrium has established itself. These various conditions ensure that poor farmers are trapped in poverty. They inherit degraded land that they cannot restore, and as a consequence remain poor and pass on an even more degraded land to their offspring. The market in such a situation would produce some equilibrium, but it would be a low-level equilibrium, penalizing the society as a whole by allowing non-efficient allocation of resources by part of its members.

In this situation, both rises and declines in prices may lead to further land degradation. When prices go down, the poor will tend to cultivate more land to maintain their income level. When prices rise, the poor, at least in the short run, will tend to use natural resources more extensively without adjusting their land management techniques in order to maximize income (as has occurred in cotton in the Sahel and cocoa in Ivory Coast). Either way, the land may be penalized by the survival strategies of farmers. It is only when the poverty cycle is broken, that higher prices may lead to more sustainable intensification of production through the use of productivity enhancing inputs like fertilizers and pesticides[2].

LAND REGENERATION IN THE LDCS: THE ROLE OF PRICES AND INCOMES

When the agriculture of a country is locked into an intergenerational tragedy of the commons, the idea that liberal market policies lead to optimal welfare no longer holds. Supportive policies become needed to induce farmers to invest in their land so that the higher-level equilibrium can be restored. In this respect, one can learn from history by looking at the developed countries. Historically, the developed industrialized countries with a strong agricultural sector have recognized the need to supplement private investment in the maintenance of land resources through direct public intervention. For the United States, Johnson and co-authors (Johnson and Timmons 1944; Johnson et al. 1947a; 1947b) argue in favour of the use of public works to improve rural livelihoods, including public investment on private-land conservation under the conditions of not providing windfall profits to private landowners. Their argument hinges on (i) equity (there were large differences between rural and urban access to social services such as education, health, water and sanitation, and general infrastructure); (ii) the provision of employment to the rural sector; and (iii) the linkages between the rural economy and the rest of the national economy, through, for example, the migration of a more qualified rural workforce to the cities. In the United States, the focus on conservation was motivated by concern over future generations, and the externalities from soil loss were viewed as contrary to the public interest. Early Acts passed by Congress in 1936 and 1956, and the more recent Food Security Act in 1985, enabled farmers to receive soil conservation payments tied to commodity supply management (Rausser 1992).

While the successful government intervention for land conservation in the industrialized countries mainly assumed the form of public investment, a typical LDC government lacks this capacity. In line with this, the literature emphasizes the importance of prices in the case of the least developed countries. Coxhead et al. (2001) argue that policy making in land use issues in the LDCs has paid too little attention to prices and markets rather than direct intervention through technology transfer, institutional innovations and other household-level actions. In their case study of the Philippines, they show that upland rice farmers in remote areas are price takers and that price shocks at the national level as well as macroeconomic instability are transmitted to them, causing alteration of land use patterns. Barbier (1990) shows that in the absence of appropriate economic incentives upland farmers on Java (Indonesia) do not adopt soil conservation practices in lieu of their traditional methods. Access to cheaper medium-term credit would make investment in terracing profitable and induce farmers to adopt these techniques, while fertilizer subsidies would encourage the cultivation of even the severely degraded soils. A time control model used by the author predicts that increasing the discount rate would enhance the use of traditional productivity boosting inputs at the expense of soil conservation inputs, while increasing the relative price of the traditionally cultivated commodities could lead to a drastic decline in soil quality over time.

In an analysis of the relation of land degradation to poverty, Grepperud (1997) delineates three kinds of production processes: (i) processes that enhance current production at the expense of long-run conservation; (ii) processes that slow down current and future degradation but at the expense of current productivity; and (iii) processes that both enhance current production and secure the environment. He shows that governments' pricing and trade policies may influence which production model is adopted.

In their study of the determinants of land conservation investment in Rwanda, Clay et al. (1998) have illustrated the key role played by prices and household wealth and liquidity. They found that non-farm income, a source of liquidity, positively affects households' conservation decisions. They also found that more stable output prices promote the use of inputs that enhance soil fertility.

In summary, the empirical evidence in the literature supports the hypothesis that a mixture of price incentives and income generation schemes could help farmers in LDCs to invest in land conservation.

Illustration from a Sahel case study

We now draw on recent results from the Sahel to give empirical support to the points raised above. The Sahel is one of the regions where land degradation poses a major threat to the future of agriculture. Burkina Faso is the Sahelian country where soil degradation seems to be the most serious. An estimated 75% of the country is suffering from important to severe degradation (Niemeijer and Mazzucato 2002). The country comprises three agro-ecological zones: the Sahelian to the north, the Sudano-sahelian in the centre, and the Sudano-guinean zone to the south-west. The first two are characterized by low agricultural potential, the third by higher potential. Annual rainfall decreases as one moves from south to north, from over 1000 mm to less than 600 mm. Moreover, the erratic nature of rainfall poses serious constraints to farmers in planning what and when to plant.

While water is the most visible constraint for agricultural production in this semi-arid setting, the condition of soils appears to be an even more limiting factor (Van Keulen and Breman 1990). The depletion of soil fertility in Burkina has been evidenced by research, through the measurement of mineral and organic balances of the soils in various zones of the country (Bikienga and Lompo 1996; Bikienga and Coulibaly 1995). Most soils are characterized by negative balances of organic matter and of the major minerals (phosphorus, nitrogen and potassium) and by a rapid deterioration of the physical structure of soils. These characteristics are typical of soils in the Sudanian zone of West Africa (Owusu-Bemoah et al. 1991). The negative mineral balances are essentially due to two factors: (i) soil erosion, which washes off soil nutrients, and (ii) the intensive and continuous cultivation of the same plots that leads to the mining of the nutrients. Nutrient mining in Burkina was estimated at 30 kg/ha of the major minerals (nitrogen, phosphorus and potassium).

Farmers try to combat soil degradation by adopting various soil conservation and improvement practices. Some of these techniques are more labour-intensive, while others are more capital-intensive. Among the latter are the commercial mineral fertilizers (NPK for example). Labour-intensive techniques commonly used in Burkina include *diguettes* (small dikes), *zaï* and mulching. Diguettes are 10-100 m long, 10-50 cm high rock barriers that collect water and hold it in the fields, prevent erosion and increase the land's water absorption capacity. Rocks are the choice materials because the bunds must be semi-permeable to prevent them from breaking under the water pressure. This labour-intensive technique (it requires 200 man-hours per hectare) is widely used in the populated semi-arid zone of Burkina. Zaï are holes of 20-30 cm in diameter and 15-20 cm deep that act as water catchments. With a spacing of 100 cm between holes, the number of holes is estimated at 20,000-25,000 per hectare of millet or sorghum field. The simple zaï can be improved by adding manure or compost in the hole, resulting in doubling or increasing yields by 50% in the short run in some regions (Ministère de l'Action Coopérative Paysanne 1990). Mulching is a simple technique of applying crop or plant residues on the land. These residues help to retain soil moisture following rainfall, and to keep temperatures down. Because of competing needs of crop residues between livestock, energy and other human uses, the technology faces a constraint in available material.

Using household survey data collected over the May-June 2004 period in 60 villages of Burkina Faso, regressions were run to assess the impacts of income on adopting these different conservation technologies and the commercial chemical fertilizers. Two sets of interesting results can be highlighted[3].

Adoption of labour-intensive, water conservation technologies. The results show that income has a positive and significant effect on the adoption of zaï in all zones, and that the relation is stronger in the Sudano-Sahelian than in the Sudano-Guinean zone. Income is also a strong determinant of the adoption of diguettes in the Sudano-Sahelian zone, but not in the Sudano-Guinean zone. Mulching, a typically cheap technology, is likewise positively related to income in the Sudano-Sahelian zone. In other terms, income is a limiting factor in all three zones for zaï, and in the Sudano-Sahelian zone for diguettes and mulching. These findings mean that if household income could be increased, investment in soil conservation could increase, especially in the degraded Sudano-Sahelian zone. However, if the hypothesis of the poverty trap is verified, households cannot undertake this additional investment because income cannot increase endogenously. This means that policy has to play a role. In the 1980s, learning lessons from the 1973 and 1984 Sahel droughts, the government in Burkina initiated actions to increase the use of soil conservation technologies in the Central plateau. Through the *Fonds de l'Eau et de l'Équipement Rural* (FEER) and other projects, equipment and food (which is a form of income in kind) were provided to village organizations to build diguettes. This contributed to boost the area covered by this erosion control technology.

Adoption of capital-intensive soil fertility management. When the use of the technology is modelled as a zero–one variable, the results suggest that income has a positive and significant effect in the Sudano-Sahelian zone, and a positive but not significant effect in the Sudano-Guinean zone. In contrast, when technology use is modelled by using household expenditure on the technology as the dependent variable, the effect of income becomes significant in the Sudano-Guinean zone and insignificant in the Sudano-Sahelian zone. These switching results have an interesting intuitive interpretation. In the Sudano-Sahelian zone, farmers are well aware of the benefits of commercial chemical fertilizer on the soil and on yields. They attempt to use it (and hence the significance of the binary variable approach), but the quantities used are so small that the relationship between expenditure and income is blurred. In contrast, farmers in the Sudano-Guinean zone are not only aware of the beneficial effect of chemical fertilizer, but they also have the means to invest in it, and richer households there spend considerable sums on fertilizer. The credit scheme provided by the cotton parastatal in this zone is a key factor in the high prevalence and level of fertilizer use. The results thus suggest that where public policy is effective, investment in land conservation may respond positively, and where incomes are low and policy is absent, there may be a potential but unmet demand for technology.

IMPLICATIONS FOR PRICE AND TRADE POLICIES

The simple neoclassical model of trade predicts that free trade provides the best environment for all trading countries, as each would be maximizing its welfare through the working of comparative advantage. Each country would be facing a given, undistorted world price that would reflect the opportunity costs of resources used in the production process. If each country then domestically tailors its own production to its own opportunity costs, given the world price, it will be exporting goods in which it enjoys a relative advantage and importing the goods for which it has relative disadvantage. The resulting situation is the maximization of total world welfare together with each individual country's welfare. The implication of this framework is that if one starts from a situation where prices are distorted by pervasive government intervention, the removal of these distortions would improve overall welfare. Each country involved in trade would benefit from the liberalization.

In spite of the resurgence of the open trade paradigm since the 1980s, many economists would accept that there are reasons why LDCs should not be exposed to a fully liberalized trade regime overnight. Some of these countries still suffer from the effects from past policies of taxing their agricultural sectors, while the developed countries were subsidizing theirs (Krueger et al. 1988). Besides, countries such as South Korea have been preparing for a liberalized trade for over three decades, while most LDCs have not. Moreover, there has been some evolution of trade theory. Non-constant returns to scale, non-homogeneous products and imperfect competition (Helpman and Krugman 1989) can be reasons for strategic government intervention, which was the path followed by a country like South Korea. Furthermore, the welfare effects indicated by most model studies of trade liberalization are only first order effects. Second order effects and long-term dynamic effects could change the conclusions significantly.

These observations suggest that LDCs should adopt a set of policies that may make them WTO-ready. This paper adds to these arguments by highlighting the intergenerational tragedy of the commons in natural resource conservation under extreme poverty. As a consequence, agricultural trade liberalization may fail to produce the intended effects for LDCs. Increased import competition as a result of trade liberalization may make domestic net sellers in these countries worse off, thereby reinforcing the poverty trap that leads to underinvestment in land. In export crops, liberalization may accelerate the over-exploitation of natural resources that is already being seen in cotton in the Sahel and cocoa in Ivory Coast. Both effects may undermine the already fragile natural resource base, compromising intergenerational equity. Policies that would induce farmers to invest in their land are needed if they are to be successful players in a liberalized global environment. Some argue (e.g. Schoenbaum 1992) that there is no conflict between free trade and domestic environmental policies, but this is only true where governments are able to protect the environment by public investment. We have already argued that LDC governments lack this capacity so that natural resource regeneration should be pursued by policies that influence markets and prices. Below we discuss possibilities for this in trade and price policies.

Trade policies

We see four areas where the LDCs may concentrate their efforts to mitigate the potential negative short-run impacts of liberalization, given their weak natural-resource base, and prepare for gains in competitiveness in the long run. These areas pertain to trading area, product choices, choice of multilateral agreements, and domestic policies.

Increasing intra-regional trade. A common phenomenon during the current wave of globalization is the emergence of regional preferential trading arrangements (PTAs). This has occurred or is occurring in all continents. Rather than being a threat to a worldwide liberalization of trade, these new schemes can be seen as permissive conditions to increased global trade. As noted by Mansfield and Milner (1999), the novelty of the new regionalism is that it involves even the most influential country in the global system, the United States[4]. These authors see the new form of regionalism as fostering liberal trade and democracy. For Perroni and Whalley (2000), the regional units can be viewed as insurance arrangements, through which the smaller countries of the units would gain access to larger, international markets that would have remained remote under bilateral trading mode.

The LDCs could reinforce their regional PTAs while remaining open to the rest of the world. For instance, many countries in Africa possess a comparative advantage in commodities imported by other African countries, and these products are often friendlier to natural resources than the products currently exported to the rest of the world. By emphasizing intra-regional trade, LDCs can alleviate the short-term negative impact stemming from the poorer quality of their natural resources, and generate income through a less aggressive use of the environment. This income could partly be reinvested in land conservation in preparation for their fuller implication in global trade.

Developing niche markets. Beyond the traditional exports (coffee, cocoa, tea, cotton), the developing world can tap into non-traditional exports that carry a higher added value and may lead to better environmental practices. These include flowers, vegetables, fruits, sesame and some processed foods. Finding niche markets in the developed countries may help poor LDCs to alleviate the disadvantage they would face due to their higher production costs of the traditional products. There are concerns however that niche market products will not benefit the small farmer. Domestic accompanying policies will be required to be tailored to the needs of the poor segments of the populations.

Entering in agreements with shared responsibility. The LDCs should favour reciprocity-based agreements in lieu of the unreciprocated agreements of the type that bind the African-Caribbean-Pacific (ACP) countries to the European Union. Although the ACP countries are lured by the short-run benefits of this type of agreement, in the long run it does not help them to become competitive on a more global scale. Promoting shared responsibility is tantamount to incorporating the full cost of PTAs in national decision making.

Domestic accompanying policies. Despite the resurgence of the paradigm of open trade from the 1980s, there is disagreement among economists as to the direct contribution of trade liberalization to economic performance. Sachs (1987) argues that the export performance of the East-Asian countries was in large part due to an active role of government in promoting exports and maintaining restrictions on imports. The example of Korea illustrates the case where a strong growth of exports (at an average annual rate of 23% between 1963 and 1990) was concomitant with a highly repressed economy characterized by substantial import tariffs and quantitative restrictions (Edwards 1993). Korea also resorted to export promotion and exchange rate policies as part of its trade liberalization strategy. For Taylor (1991), there are "no great benefits (plus some loss) in following open trade and capital market strategies". He argues therefore that internally based development policies will be the best choice for developing countries.

Price policies

Agricultural prices will play a major role if LDCs are to draw sustainable benefits from global liberalization. These countries are in large majorities composed of farm households whose incomes mostly depend for 50% or more on agriculture. As argued above, income is directly correlated with investment in land conservation. Because the latter is a major determinant of productivity and agricultural income is the product of prices and quantities (less costs), prices have an important influence on agricultural productivity. Using data on 18 developing countries, Fulginiti and Perrin (1993) have established a significant positive relationship between past prices and current productivity levels of agricultural resources. Although these are somewhat outdated, the relationship is probably still relevant.

A common practice that has shaped the past of the developing countries has been the taxation of agriculture, in particular export crops[5]. Fulginiti and Perrin (1993) estimated that taxation of agriculture had caused a 26% loss in productivity in the countries they studied. Although the bulk of agricultural taxation has been eliminated during the 1980s and 1990s in most LDCs following the implementation of adjustment programs, implicit or explicit taxes on some export crops and fertilizer still remain in some African countries. This is the case of the integrated cotton production approach in some of the French-speaking franc zone countries, with large cotton companies acting as intermediaries in input supply and product purchase through contractual arrangements and exchange rate overvaluation as a source of implicit taxation. At the turn of the present millennium, it was estimated that as little as 35% of a price increase on the international market of cotton was passed on to the producer in Burkina Faso (Sirima and Savadogo 2001). Townsend (1999) assessed the price policies of African countries in the areas of export crops, food crops, fertilizer and macroeconomic policy for the period 1996-97. For export crops, Burkina Faso, Benin, Ghana, Togo and Mali ranked very poorly, with the transmission of international price changes below the expected levels.

Producer prices of agricultural products in some LDCs are also low because of structural factors, including poor roads that hinder market integration. The high transaction costs that result from poor transport systems lead to prices being low in surplus production areas and high in deficit areas. This situation acts as an indirect tax on producers. Countries need to integrate infrastructural investment as part of a policy package toward a liberalized agriculture.

Finally, there is a role for the international community. Depressed international prices may accelerate natural resource exploitation in the LDCs as argued above. A positive discrimination towards the LDCs (without, however, the non-reciprocal arrangement of the EU-ACP type) would be a solution. As Resnick (2004) shows on the basis of World Bank data for 2003, in the United States alone, the subsidies under the Farm Bill total $15-20 billion per year, more than the value of Africa's total annual agricultural export. One also notes that the share of official development assistance to Africa from the European Union and the United States that is allocated to agriculture has been on a steady decline, from 14-16% in 1990 to 6-8% in 2002. A reactivation of external assistance under some form that would act as a subsidy to the small farmers in the poorest of the LDCs should be considered.

SUMMARY AND CONCLUSION

The objective of this paper was to draw attention to limitations to the neoclassical tenet that trade liberalization in agriculture will benefit all trading partners. The paper looked at the particular case of investing in land regeneration and argued that this investment possessed some public good characteristics when looked at from an intergenerational standpoint and in the context of widespread and deep poverty. As a consequence, private investment in regeneration may lie below the socially optimal level. Without proper intervention, farmers in LDCs will be facing a weak resource base and fail to be competitive in a global trade environment.

Three sets of conclusions may be drawn. First, LDCs will need to enact appropriate trade policies in order to (i) prevent the potential negative impact of a liberalized global trade on natural resources, and (ii) to ensure that they can stand to benefit from liberalization in the long run. Such policies include the formation of regional PTAs, the tapping of niche markets, the abandonment of unreciprocated international arrangements for reciprocity-based agreements, and a set of domestic policies friendly to trade. Second, LDCs will need to reconsider their price policies. The implicit taxation of agricultural output prices that still prevails in some countries needs to be discontinued. Structural development (e.g., better road infrastructure) is also needed to lower the high transaction costs in trade that tend to depress producer prices. Finally, the international community will need to develop special programs to assist the small farmer in the poorest countries.

NOTES

1. Intertemporal externalities are similar to spatial externalities as both involve the motion from one point to another. The difference is that space is reversible while time is not, but memory looping allows simulating time reversibility.
2. Unsustainable expansion of cultivation as a response to price rises is the consequence of a situation of poverty and has a short-run nature. If, through some policy, the poverty cycle can be broken, then the rational and expected response of high international prices will be the intensification of production through the use of productivity-enhancing inputs (fertilizers, pesticides). However, many cash-crop producers in the poorer countries still have very small operating sizes. For the particular case of Burkina Faso, a small country but a large cotton producer on the African scale, the mean farm size among cotton producers in 2003 is only 8.4 ha, the median is 6.3 ha, while the largest farm is 37.7 ha (sample data from the Bâle province, a major cotton-producing zone of Burkina (Bambio 2006)). About half of the total farm is allocated to cotton. At such levels, most internationally traded cash-crop producers barely escape poverty, and as long as people remain poor, the expected rational response will appear to be a long shot.
3. For details on the model and data, see Savadogo (2004).
4. The North American Free Trade Agreement (NAFTA) includes the United States, Canada and Mexico.
5. In the sample used by Fulginiti and Perrin, the nominal protection rates varied from -13% (Brazil) to -53% (Ivory Coast, Egypt and Zambia). The data covered the period of the 1960s to the early 1980s. South Korea was part of the sample and had a positive protection rate of +16%.

REFERENCES

Bambio, Y., 2006. *Contrainte de crédit et productivité agricole en zone cotonnière du Burkina Faso.* University of Ouagadougou, Ouagadougou. Doctoral thesis. University of Ouagadougou

Barbier, E.B., 1990. The farm-level economics of soil conservation: the uplands of Java. *Land Economics,* 66 (2), 199-211.

Bikienga, I.M. and Coulibaly, O., 1995. *Fertilisation et intensification agricole au Burkina Faso: paper presented at Colloque International sur l'Intensification Agricole au Sahel, Bamako, November 28-December 2.* Ministère de l'Agriculture et des Ressources Animales, Burkina Faso.

Bikienga, I.M. and Lompo, F., 1996. *Développement d'une stratégie nationale de gestion de la fertilité des sols au Burkina Faso: paper presented at seminar 'Lier la gestion de la fertilité des sols au développement du marché des intrants et des produits agricoles pour une agriculture durable en Afrique de l'Ouest,' Lomé, Togo, November 19-22.*

Clay, D., Reardon, T. and Kangasniemi, J., 1998. Sustainable intensification in the Highland Tropics: Rwandan farmers' investments in land conservation and soil fertility. *Economic Development and Cultural Change,* 46 (2), 351-377.

Coxhead, I., Rola, A. and Kim, K., 2001. How do national markets and price policies affect land use at the forest margin? Evidence from the Philippines. *Land Economics,* 77 (2), 250-267.

Dregne, H.E., 1990. Erosion and soil productivity in Africa. *Journal of Soil and Water Conservation,* 45 (4), 431-436.

Dregne, H.E. and Chou, N.T., 1992. Global desertification dimensions and costs. *In:* Dregne, H.E. ed. *Degradation and restoration of arid lands.* Texas Tech University, Lubbock. [http://www.ciesin.org/docs/002-186/002-186.html]

Edwards, S., 1993. Openness, trade liberalization, and growth in developing countries. *Journal of Economic Literature,* 31 (3), 1358-1393.

El-Beltagy, A., 1997. *Land degradation: a global and regional problem.* ICARDA, Aleppo. [http://www.unu.edu/millennium/el-beltagy.pdf#search=%22land%20degradation%20global%20regional%20problem%22]

Eswaran, H., Lal, R. and Reich, P.F., 2001. Land degradation: an overview. *In:* Bridges, E.M., Hannam, I.D., Oldeman, L.R., et al. eds. *Responses to land degradation: proceedings 2nd international conference on land degradation and desertification, Khon Kaen, Thailand.* Oxford Press, New Delhi. [http://soils.usda.gov/use/worldsoils/papers/land-degradation-overview.html]

Fulginiti, L.E. and Perrin, R.K., 1993. Prices and productivity in agriculture. *The Review of Economics and Statistics,* 75 (3), 471-482.

Grepperud, S., 1997. Poverty, land degradation and climatic uncertainty. *Oxford Economic Papers,* 49 (4), 586-608.

Gretton, P. and Salma, U., 1997. Land degradation: links to agricultural output and profitability. *The Australian Journal of Agricultural and Resource Economics,* 41 (2), 209-225.

Helpman, E. and Krugman, P.R., 1989. *Trade policy and market structure.* MIT Press, Cambridge.

Holden, S., 1997. *Environmental problems and agricultural development in the less developed countries.* Norwegian University of Life Sciences, Aas. Discussion Paper no. D-2/1997.

Johnson, V.W. and Timmons, J.F., 1944. Public works on private land. *Journal of Farm Economics,* 26 (4), 665-684.

Johnson, V.W., Timmons, J.F. and Howenstine Jr., E.J., 1947a. Rural public works. Part I: Needed improvements and useful jobs. *The Journal of Land and Public Utility Economics,* 23 (1), 12-21.

Johnson, V.W., Timmons, J.F. and Howenstine Jr., E.J., 1947b. Rural public works. Part II: Building a positive program. *The Journal of Land and Public Utility Economics,* 23 (2), 132-141.

Jongeneel, R. and Koning, N., 1999. Unjustified claims of welfare economics: potential Pareto improvement revisited. *Tijdschrift voor Sociaalwetenschappelijk Onderzoek van de Landbouw,* 14 (3), 114-126.

Krueger, A.O., Schiff, M. and Valdés, A., 1988. Agricultural incentives in developing countries: measuring the effect of sectoral and economywide policies. *The World Bank Economic Review,* 2 (3), 255-271.

Lal, R., 1995. Erosion-crop productivity relationships for soils of Africa. *Soil Science Society of America journal,* 59, 661-667.

Mansfield, E.D. and Milner, H.V., 1999. The new wave of regionalism. *International Organization,* 53 (3), 589–627.

Ministère de l'Action Coopérative Paysanne, 1990. *Rapport de fin de campagne 1989/90 du projet expérimental zaï.* Ministère de l'Action Coopérative Paysanne, Ouagadougou.

Narayanan, S. and Gulati, A., 2002. *Globalization and the smallholders: a review of issues, approaches, and implications.* IFPRI, Washington. Market and Structural Studies Division Discussion Paper no. 50. [http://www.ifpri.org/divs/mtid/dp/papers/mssdp50.pdf]

Niemeijer, D. and Mazzucato, V., 2002. Soil degradation in the West African Sahel: how serious is it? *Environment (Washington DC),* 44 (3), 20-31.

Owusu-Bemoah, E., Acquaye, D.K. and Abekoe, M., 1991. Efficient fertilizer use for increased crop production: use of phosphorus fertilizers in concretional soils of Northern Ghana. *In:* Mokwunye, A.U. ed. *Alleviating soil fertility constraints to increased crop production in West Africa.* Kluwer, Dordrecht.

Perroni, C. and Whalley, J., 2000. The new regionalism: trade liberalization or insurance? *The Canadian Journal of Economics,* 33 (1), 1-24.

Rausser, G.C., 1992. Predatory versus productive government: the case of US agricultural policies. *The Journal of Economic Perspectives,* 6 (3), 133-157.

Resnick, D., 2004. *Smallholder African agriculture: progress and problems in confronting hunger and poverty.* International Food Policy Research Institute, Washington. Development Strategy and Governance Division Discussion Paper no. 9. [http://www.ifpri.org/divs/dsgd/dp/papers/dsgdp09.pdf]

Sachs, J.D., 1987. Trade and exchange rate policies in growth-oriented adjustment programs. *In:* Corbo, V., Goldstein, M. and Khan, M. eds. *Growth oriented adjustment programs.* IMF, Washington.

Savadogo, K., 2004. *Public goods, natural resources and equity considerations in assessing the impacts of agricultural trade liberalization in the LDCs: application to the Sahel: paper presented at the H.E. Babcock Workshop on "Agricultural trade liberalization and the least developed countries: how should they respond to developments in the WTO?" Wageningen (The Netherlands), 2-3 December 2004.*

Scherr, S.J. and Yadav, S., 1997. *Land degradation in the developing world: issues and policy options for 2020.* IFPRI, Washington. IFPRI 2020 Brief no. 44.

Schoenbaum, D.J., 1992. Free international trade and protection of the environment: irreconcilable conflict? *The American Journal of International Law,* 86 (4), 700-727.

Sirima, B. and Savadogo, P., 2001. *Burkina Faso: competitiveness and economic growth: policies, strategies, actions.* Ministry of Economy and Finance, Ouagadoguou.

Stocking, M. and Murnaghan, N., 2001. *Handbook for the field assessment of land degradation*. Earthscan Publications, London.

Taylor, L., 1991. Economic openness: problems to the century's end. *In:* Banuri, T. ed. *Economic liberalization: no panacea*. Clarendon, Oxford.

Townsend, R.F., 1999. *Agricultural incentives in Sub-Saharan Africa: policy challenges*. World Bank, Washington, DC. World Bank Technical Paper no. 444. [http://www-wds.worldbank.org/servlet/WDSContentServer/WDSP/IB/1999/09/25/000094946_99090805303990/Rendered/PDF/multi_page.pdf]

Van Keulen, H. and Breman, H., 1990. Agricultural development in the West African Sahelian region: a cure against land hunger? *Agriculture, Ecosystems and Environment*, 32 (3/4), 177-197.

CHAPTER 6

TRADE LIBERALIZATION IN COTTON AND SUGAR

Impacts on developing countries

ANDRÉ MELONI NASSAR

General manager, Institute for International Trade Negotiations (ICONE), São Paulo, Brazil

INTRODUCTION

The past five years have seen a shift in the regulation of the international markets of agricultural products. For the first time since the settlement of the Agricultural Agreement of the Uruguay Round, agricultural policies of developed countries have been questioned in the World Trade Organization (WTO).

In 2002, Brazil, Australia and Thailand contested the European Communities' export subsidies on sugar. In the same year, Brazil contested the export credits and domestic subsidies provided by the U.S. government to its cotton producers. In both cases the reports of the Panel and the Appellate Body were in favour of the complainants.

Other developing countries reacted differently to these cases. The ACP countries were sceptical about the sugar case, which threatened to erode their preferential access to the European market. On the other hand, the cotton case was followed with high expectations by the West African cotton-exporting countries.

At the moment that this chapter was being concluded, the reform of the European Union sugar regime and of the United States cotton policies was still a subject of discussion. The EU approved the reform of its sugar policy in 2006 to meet the recommendations of the sugar panel, but it still has to be implemented. Because the new intervention prices remain higher than world prices, the EU will not be allowed to export more than its Uruguay Round commitment of 1.2 million tons per year. Any larger quantity will be taken as an indication that the cross-subsidization of exports due to the domestic subsidies has not been eliminated. This

N. Koning and P. Pinstrup-Andersen (eds.), Agricultural Trade Liberalization and the Least Developed Countries, 83–103.

makes it important to evaluate how the production and exports respond to implementation of the reform.

The US, for its part, has decided to partially implement the recommendations of the cotton panel. Export subsidies policies have been reformed – in the case of export guarantee programs – or eliminated – in the case of the so-called Step 2 subsidies. However, domestic subsidies that cause adverse effects on the international market have remained unchanged. One of the reasons for this is that reforming domestic policies for cotton would necessitate a reform of all commodity programs. The existing Farm Bill will be reviewed in 2007 and the U.S. government has decided not to pre-empt the debate.

In addition to the panel recommendations, the multilateral negotiations of the Doha Round can also lead to the liberalization of the cotton and sugar markets. The 2001 Doha Mandate called for the elimination of export subsidies, significant reductions in domestic support, and substantial improvements in market access. In 2004, cotton was given a special status in the negotiations following a joint proposal by Benin, Burkina Faso, Chad and Mali that followed the same motivations as those of Brazil in the cotton case. Since the suspension in the negotiations in July 2006, however, the course of the WTO Doha Round has become uncertain.

This paper discusses the WTO process in sugar and cotton and its potential impact on different developing countries. The next section discusses the structure of the cotton and sugar world markets as well as the role of the developing countries in each of them. Section *"How developed countries' policies affect developing countries"* analyses how cotton and sugar world markets are distorted. This section focuses on the impact of developed countries' policies in the market. Section *"Liberalization of the world markets for cotton and sugar: impacts on developing countries"* is dedicated to discussing different elements of market liberalization: the forces both in favour of it and against it, scenarios taking into account the implementation of the WTO cases and Doha Round negotiations, and the balance between winners and losers. This section will also present a review of papers discussing the benefits of liberalization of the sugar and cotton markets. Section *"Final remarks"* contains concluding remarks.

INTERNATIONAL MARKETS OF COTTON AND SUGAR

Cotton

Cotton is traded internationally in bales. Once harvested, the lint is separated from the seed (ginning process) before being sold to spinning mills. Although some spinners source cotton directly from ginning companies in exporting countries, the typical exportation transaction is performed by trading companies. Cotton lint is classified according to the quality of the fibre. The quality of the cotton is measured by staple length, strength, colour, uniformity, foreign matter and stickiness. These characteristics may vary depending on suppliers and crop year. The seed variety and the technology used for ginning are the main determinants of these characteristics.

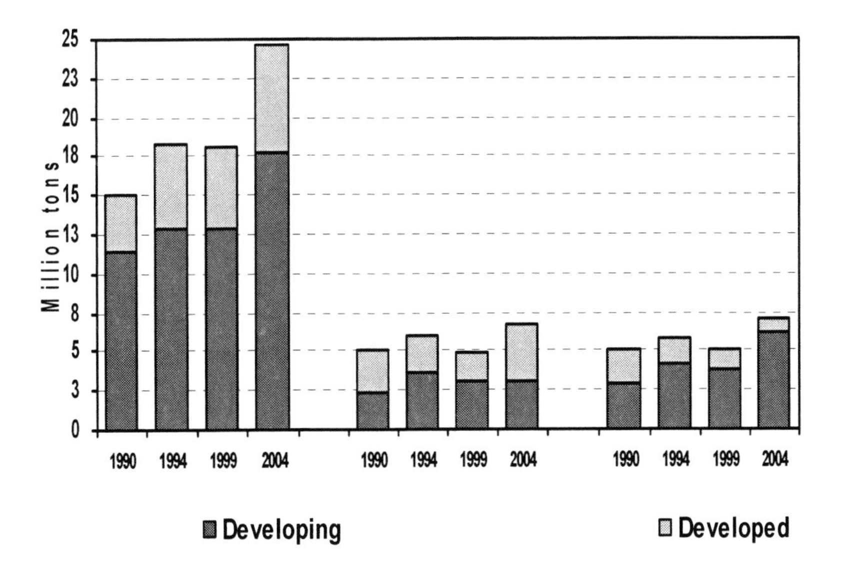

Source: FAOSTAT (excluding intra-EU trade)

Figure 1*. Production, exports and imports of cotton by developing countries and developed countries*

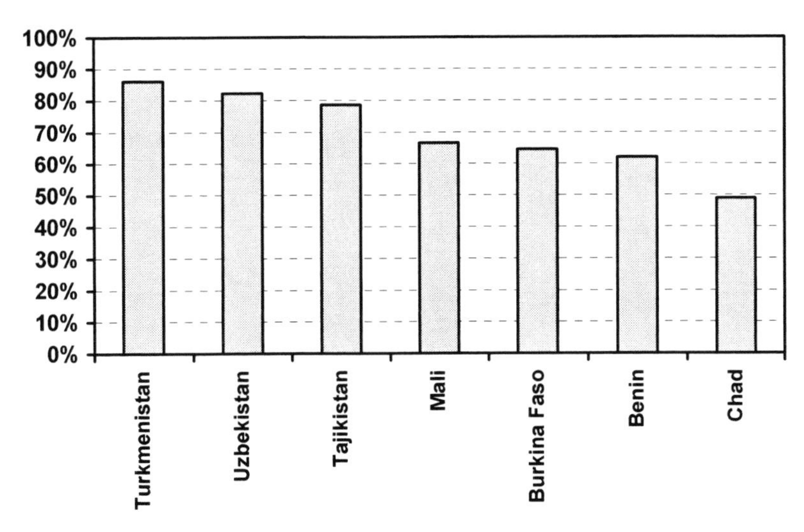

Source: FAOSTAT

Figure 2*. Share of the cotton exports on total agricultural exports*

Developing countries are responsible for the majority of the production and imports (Figure 1). The U.S. is the leading exporter, with 40 percent of world exports. For some countries, cotton is the major export product. In Uzbekistan – the second-largest world exporter – 82% of total exports are concentrated in cotton. In Mali, Burkina Faso and Benin – three of the four countries that supported the sectoral initiative – cotton accounts for more than 60% of total agricultural exports (Figure 2).

World cotton prices fell between 1995 and 2002 (Figure 3) because of two major factors. One was the stabilization of world imports due to a reduction of import demand from China. The other was the expansion of U.S. exports, started in 1999. The American subsidies for cotton helped U.S. cotton producers to increase their production despite the non-dynamic world market. This exacerbated the international price fall and was the main motivation for the WTO cotton case and the African cotton initiative.

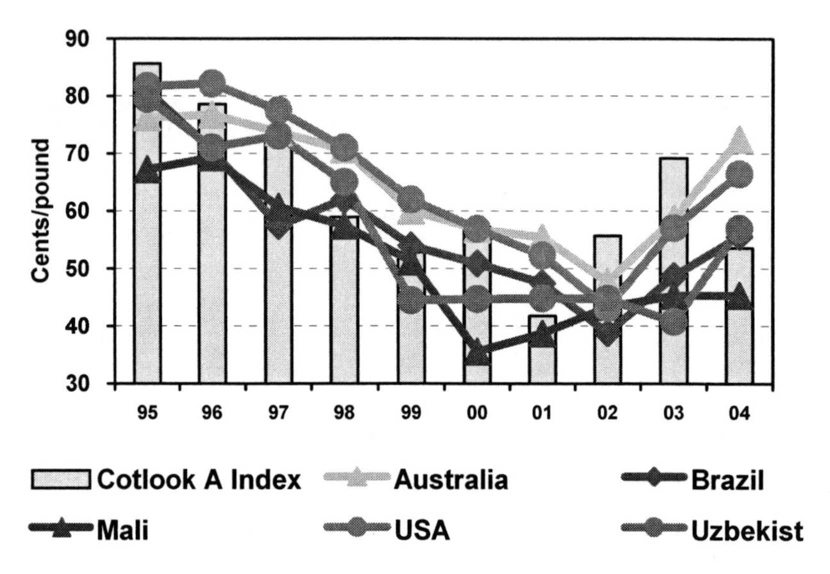

Source: ICAC, FAOSTAT

***Figure 3**. World price of cotton and export unit values*

In the last four years, world cotton trade has shown a consistent growth. From 1.5 percent in the nineties, the average annual growth rate jumped to 5.3 percent from 2000. The main cause was the ending of the WTO Agreement on Textiles and Clothing in 2004. Companies started to make new investments, even some years before the agreement had officially ended, thereby stimulating new demand. China, after some years out of the market, started to import huge volumes in 2002. Imports also increased in Pakistan, Turkey, Thailand and several other developing countries. The demand from the Indian textile industry likewise increased, but imports decreased due to a recovery in domestic cotton production. On the other hand, in a country like Mexico imports decreased as soon as its textile industry was no longer protected by the Textile Agreement (Figure 4).

Production in developing countries responded to the increased demand. From 2000 to 2004, their cotton production grew 5.6 percent a year. However, the performance of different countries varied substantially. Since 2002, the production of traditional cotton producers such as Turkey, Egypt and Uzbekistan has remained mostly constant despite the upturn in world production. Conversely, Brazil has appeared as a significant producer, while Sub-Saharan African production has grown 5 percent annually in the last five years. These countries, and some others like Paraguay and Egypt, are expanding their exports. Sub-Sahara Africa's export growth rate even reached 6.6 percent at one point. However, this region accounts for only 16 percent of world exports. However, by far the greatest increase in exports occurred not in developing countries but in the United States, where the combination of a rapid increase in production and decreasing demand from the U.S. textile industry resulted in a 76 percent increase in exports from 2000.

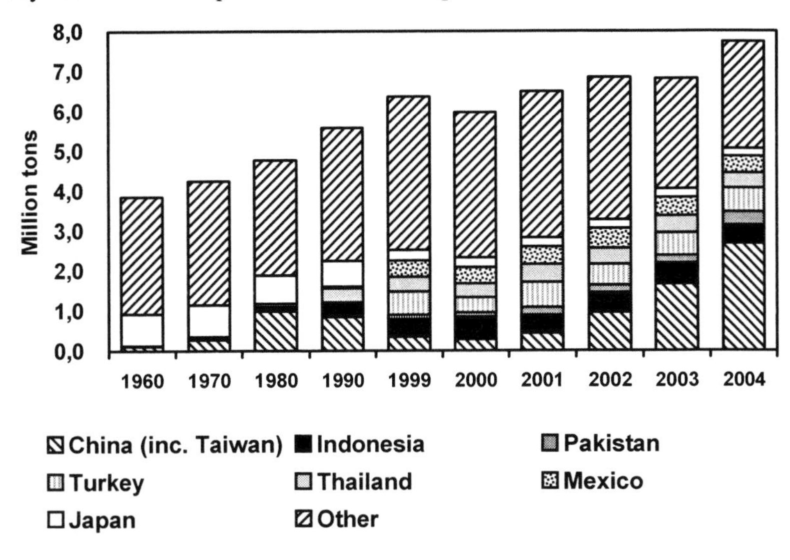

Source: ICAC

Figure 4. *Main importers of cotton*

Sugar

Sugar is traded internationally in raw and white forms. According to USDA data, raw sugar accounts for 56 percent of the world sugar market and white sugar for 44 percent[1]. Raw sugar, which is traded in bulk, originates from sugar cane, a typical product of tropical areas. It is produced from the fermentation of the cane juice and the raw material for the refineries. Many countries import raw sugar and refine it domestically. Beet sugar is produced in temperate areas and is necessarily traded as white sugar (in sacks). While sugarcane producers may export both raw and white sugar, beet sugar producers export only white sugar.

Because of climate conditions, most countries are specialized in either cane or beet sugar. While 73.5 percent of beet sugar is produced in developed countries, 91.4 percent of cane sugar is produced in developing countries. The United States is an exception, given that it cultivates as much beet sugar as cane sugar.

World production of sugar is around 142 million tons. Around 47 million tons per year is traded internationally[2]. The ratio of trade to production has increased from 0.26 percent in the beginning of the 1990s to 0.33 in recent years.

Among the developed countries, the EU dominates the white sugar market. From the 7.1 million tons of developed country exports in 2004/05, the EU-25 exported around 6 million tons (86 percent). In the raw sugar market, Australia accounts for 99 percent of the total exports of developed countries.

Developing countries sell 13.6 million tons of white sugar abroad (Figure 5). Brazil, Thailand, India, Colombia, South Africa, Saudi Arabia and the United Arab Emirates account for 70 percent of these exports[3]. The difference between developed countries and developing countries is still stronger in the raw sugar market, where developing countries export 22.8 million tons. Brazil, Thailand, Colombia, South Africa, Saudi Arabia, Guatemala and Cuba account for 81 percent of these exports.

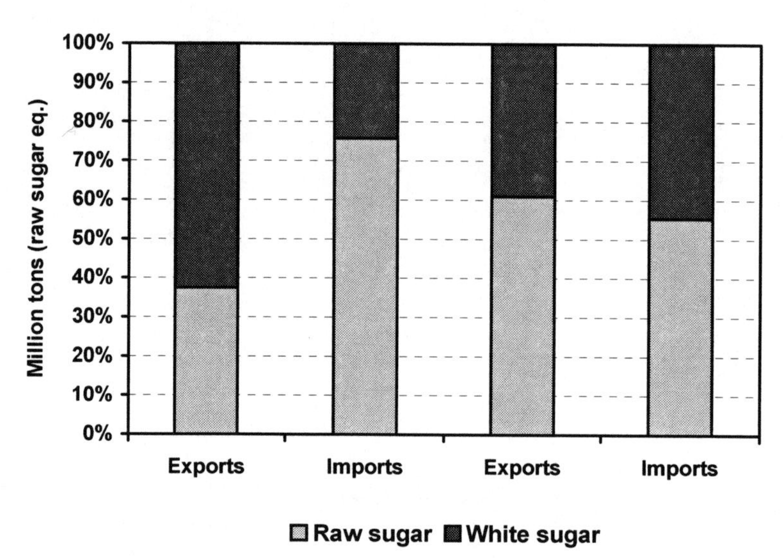

Source: USDA/PSD

Figure 5*. Trade of raw and white sugar: performance of developing and developed countries (2003)*

Developing countries are also the most important importers. They import 35.3 million tons, compared with around 9.8 million tons imported by developed

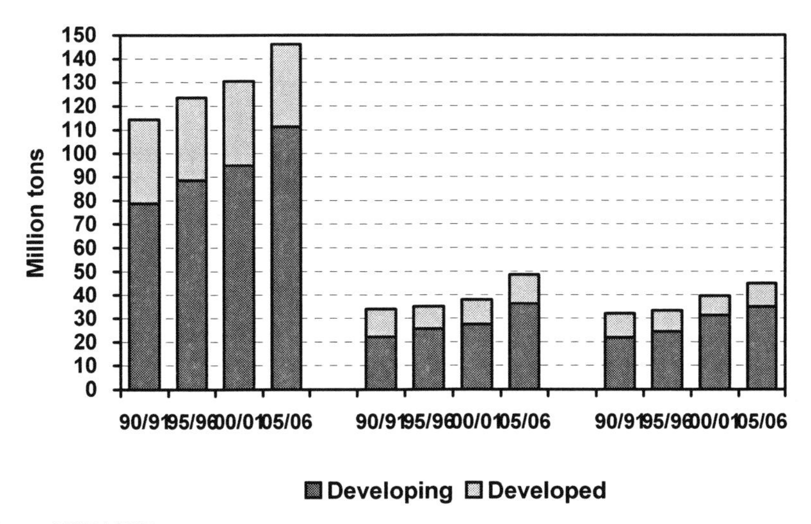

Source: USDA/PSD

Figure 6. *Production, exports and imports of sugar by developing and developed countries*

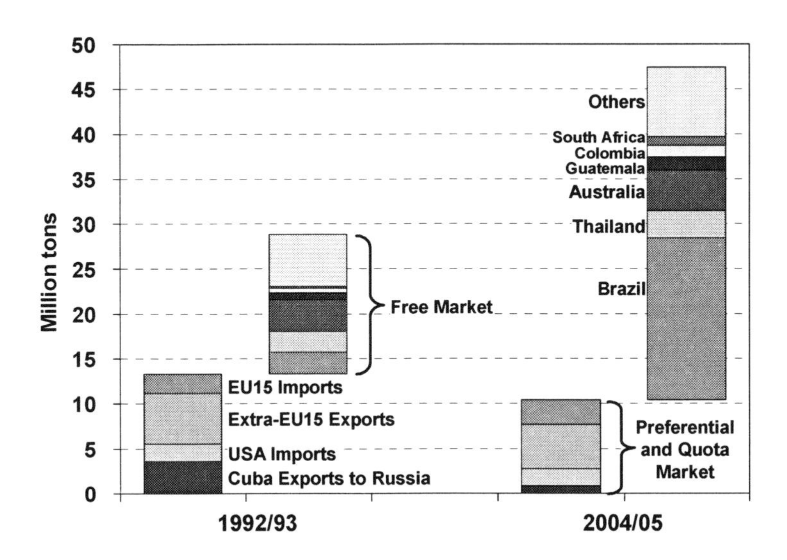

Source: USDA/PSD; European Commission/COMEXT

Figure 7. *Comparison between the free market and the regulated market of sugar* countries. While purchases by developed countries are concentrated in the raw sugar market (75 percent of their total imports), those of developing imports are more balanced between raw sugar (60 percent) and white sugar (40 percent).

The historical data highlight two important facts. In the first place, developing countries are becoming more important in production, exports and imports (Figure 6) In the second place, with the growth of the volume traded internationally, the importance of the free market over the regulated (preferential and quota) market has increased (Figure 7)[4].

HOW DEVELOPED COUNTRIES' POLICIES AFFECT DEVELOPING COUNTRIES

Cotton

Although import restrictions are less important in cotton, the world cotton market is strongly distorted by price-based payments and export competition policies (ICAC 2002; FAO 2004). The three countries that give most support to their cotton producers are the U.S., the EU and China (Figure 8).

The cotton program of the European Union keeps its net imports lower than they otherwise would be (Karagiannis 2004). Cotton producers are supported by guaranteed prices and direct payments. The aggregate support level is determined by the difference between the indicative (guaranteed) price (€ 1,063/ton) and the world price as well as the proportion of production to the maximum guaranteed quantity (782,000 tons for Greece and 249,000 tons for Spain). An excess of production over

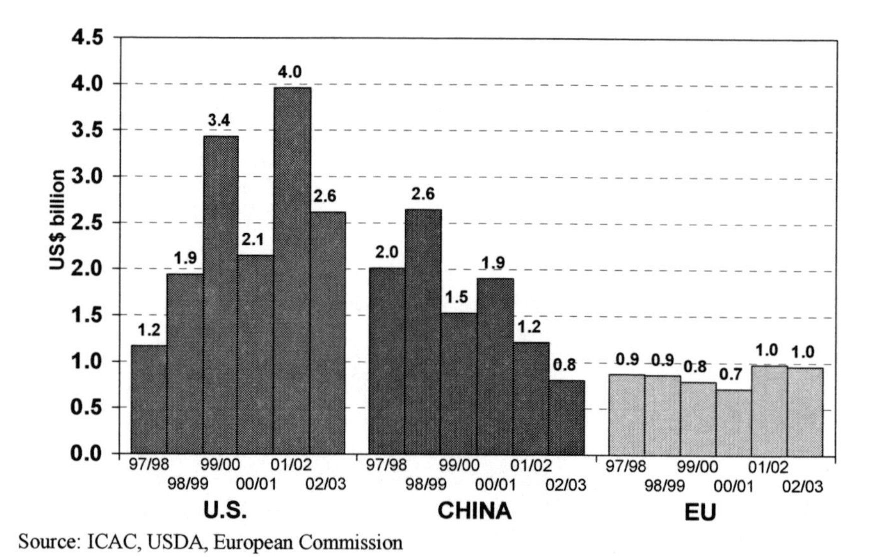

Source: ICAC, USDA, European Commission

Figure 8*. Amount of support given by the governments to the cotton farmers*

this quantity leads to a discount on the indicative price so that the total amount of support is reduced. The discount has been consistently higher in Greece than in Spain because the surplus of production in the former was higher. Chinese cotton production has been traditionally supported by price supports and subsidies for transportation, marketing and public stockholding. However, reform in 1999 reduced the level of support. Besides, China's accession to the WTO resulted in a reduction in government intervention, and a further adjustment of policies to WTO rules is to follow. China has a tariff rate quota of 894,000 tons on which it imposes a one-percent tariff against an over-quota tariff of 40 percent. Since 2003, the quota volume has been extended to meet increasing domestic demand requirements. To sustain domestic prices, a state trade enterprise operates one-third of total imports. In the future, China will probably shift to direct payments to producers.

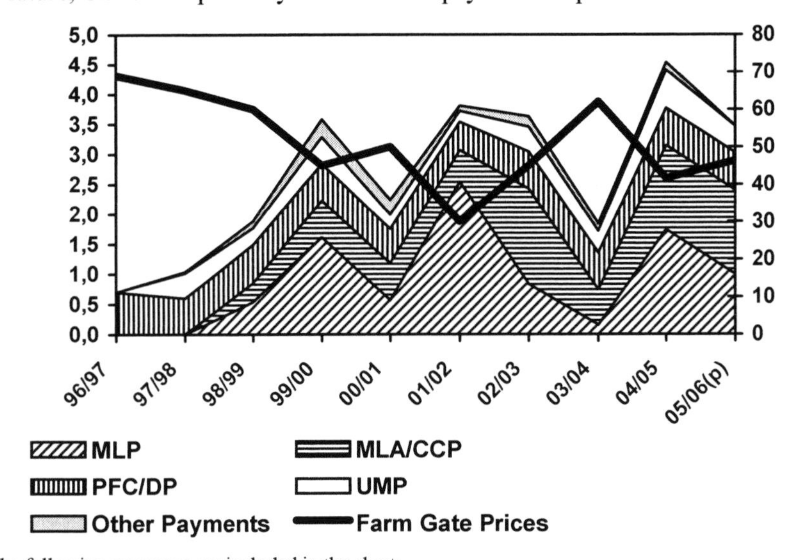

The following programs are included in the chart:

1. MLP (Marketing Loan Programs): LDP (Loan Deficiency Payments), MAL (Marketing Assistance Loans) and CEG (Certificate Exchange Gains).
2. MLA (Marketing Loan Assistance) and CCP (Certificate Exchange Gains).
3. Decoupled payments: PFC (Production Flexibility Contracts) and DP (Direct Payments).
4. UMP (User Marketing Payments – Step 2).
5. Other Payments: CSP (Cotton Seed Payments), Storage payments and Commodity Loan Interest Subsidy and Fees/Levies.

Source: USDA-CCC & WTO. Elaboration: ICONE

Figure 9. *US: Domestic support to cotton*

It should be noted that the cotton policies of the EU and China have limited distortive effects. The EU is not a major player in the world market, while China's cotton policy has become less distortive since this country has reduced its cotton support and increased its imports. The world cotton market is mainly affected by U.S. policies. U.S. cotton programs shield cotton producers against price fluctuations and support producer revenue in ways that have a significant negative effect on world prices (Gillson et al. 2004; Goreux 2004).

According to the USDA, the U.S. government transferred US$ 4.5 billion in 2004/05 to its cotton producers, in other words US$ 180,000 per farmer. Cotton farmers receive subsidies from different programs (Figure 9). Marketing Loan Programs (MLP) are tied to production and inversely related to world prices[5]. As a consequence, they strongly affect production decisions made by cotton farmers.

Counter-cyclical payments (CCPs) are likewise related to world prices, but tied to fixed past production[6]. They were created by the 2002 Farm Bill to replace the Marketing Loss Assistance (MLA) that was introduced as an emergency measure when the programs of the 1996 Farm Bill proved unable to prevent strong decreases in farm income in the period of low prices between 1998 and 2002. The institutionalisation of this assistance in the form of the CCPs is a clear indication that U.S. cotton farmers are becoming more dependent on government transfers. CCPs affect producers' decisions because they reduce price risks, and because the 2002 Farm Bill has allowed producers to update the base acreage on which these payments are based.

Something similar is true for the direct payments (DPs), which are linked to the CCP base acreage. Although they are not tied to current production and prices, they represent an additional income support, even in periods of high prices.

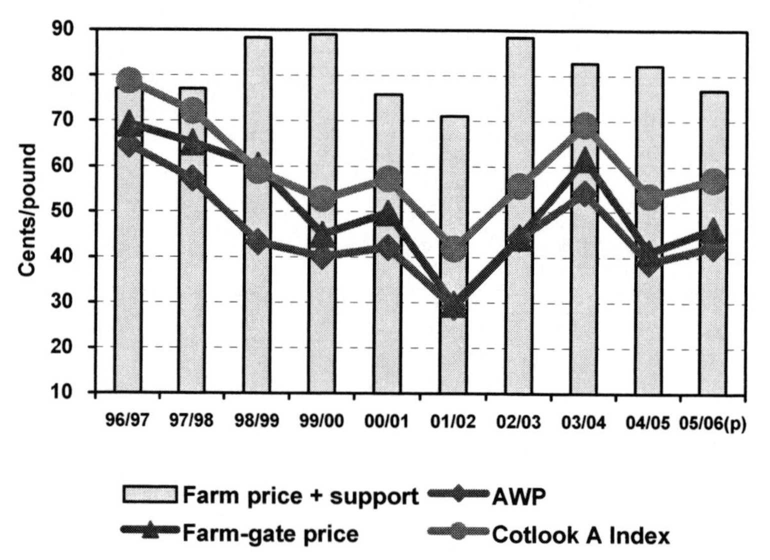

Source: USDA-CCC & WTO. Elaboration: ICONE

***Figure 10**. Comparison between the prices received by US farmers and the world prices*

In addition to these payments, which are directly or indirectly linked to production, the 1996 FAIR Act also introduced another support instrument for the cotton sector that was more specifically linked to exports. This was the User Marketing Payment (UMP) or Step 2 payment[7]. This payment is notified under the Amber Box. Expenditures on it reached their peak in 1999 (US$446 millions).

The recent evolution of U.S. production and exports of cotton clearly exhibits the influence of these various support measures (Baffes 2003; 2004). Although domestic demand was decreasing, production continued to grow rapidly, even in the period of falling international prices. Evidently, the support measures eliminated the incentives for farmers to adjust their production to the new conditions in the domestic and international markets.

Also, there is a strong relation between the evolution of U.S. exports of cotton and the low prices between 1998/99 and 2001/02. While Sub-Saharan African exports stopped growing, and Uzbekistan exports decreased during these years, U.S. exports recovered quickly (Watkins 2002). Figure 10 shows that the actual price decrease (farm-gate price plus government payments) was less sharp for the U.S. cotton farmers. In 2001/02, when international prices were at their lowest levels, U.S. farmers decided to plant more cotton (Figure 11). The peak of U.S. production in that year coincided with the peak of transfers from the U.S. government to cotton farmers.

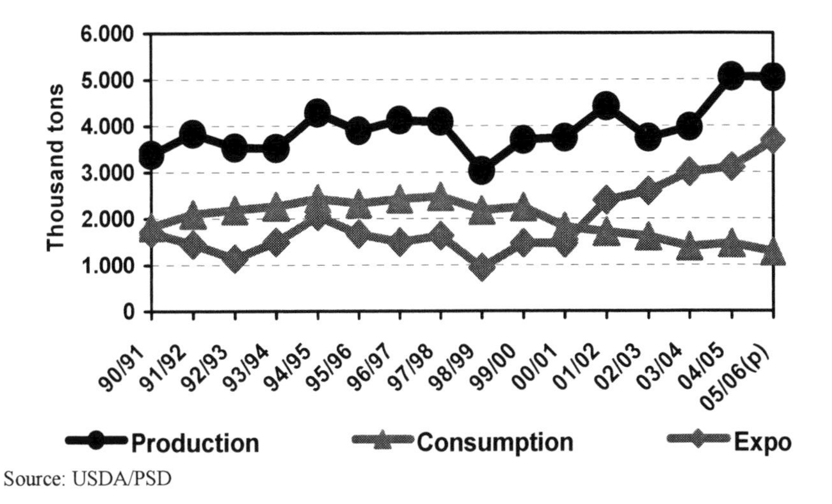

Source: USDA/PSD

***Figure 11**. U.S. cotton supply and demand*

The final report of the cotton panel found that the user marketing payments, together with the export credits programs (GSM-102, GSM-103 and SCGP), should be considered export subsidies. Given that the U.S. had no export subsidy commitments for cotton, they were illegal and should be eliminated or reformed in order to eliminate the export subsidy component. In August 2006, the UMP was repealed and the export guarantee programs have been reformed in line with the panel's recommendations[8].

The panel report considered the DPs non trade-distorting. However, it found them product-specific and therefore not notifiable as a green box provision. As for the MLPs and CCPs, the report found that these subsidies could cause price depression and serious disadvantages for cotton exporters. The U.S. should therefore modify these programs in order to eliminate their price-distorting effects. As yet, no modification to the provisions of these programs has been made by the U.S. government. However, the panel results are strongly influencing the negotiations for the new farm bill[9].

Sugar

The world sugar market is being distorted by trade barriers and export subsidies (Dymock 2002), which keep domestic prices in important OECD countries at high levels (Figure 12). According to the CIE (2002), domestic sugar prices in the U.S., the EU and Japan are 153 percent, 211 percent and 383 percent of international prices, respectively.

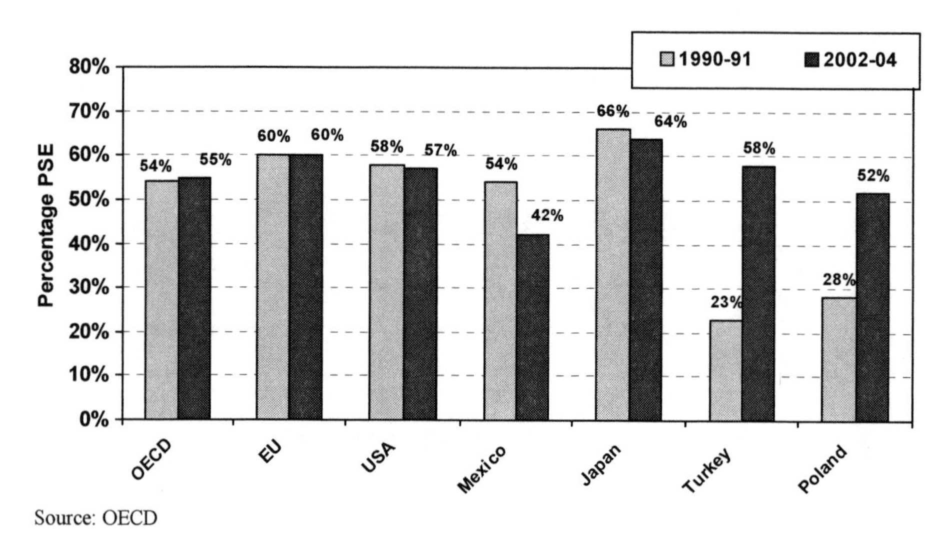

Source: OECD

***Figure 12**. Sugar producer support estimate (PSE as ratio of gross farm receipts)*

Japan and the U.S. are net importers of sugar and use border mechanisms to protect domestic production. The U.S. has three WTO tariff rate quotas and one quota for the intra-NAFTA trade. Quota volumes are allocated to exporting countries according to the export performance in the period from 1975 to 1981[10]. Over-quota imports are prevented through prohibitive tariffs and entry prices. The domestic market is regulated by minimum prices. If domestic market prices fall below these, sugar producers are allowed to sell their production to the USDA. Japan supports its sugar industry by importing raw sugar at international prices and reselling it domestically at higher prices. Thus the domestic prices are kept above the minimum prices set by the government. A semi-governmental monopoly (Agriculture and Livestock Industries Corporation – ALIC) manages the internal market and sets the domestic prices. Both the U.S. and Japan distort the market, not by subsidizing exports but by importing less than they would if their markets had been open.

The EU sugar policy is much more complex. Like the U.S. and Japan, the EU imposes high border protection. Prohibitive over-quota tariffs (€419.0 per ton for white sugar and €339.0 for raw sugar) and additional duties under a special safeguard clause that are to be applied when the representative price falls below a trigger price, prevent any out-of-quota imports. However, the EU also has a surplus of sugar, which is exported with export subsidies. Moreover, the EU has an additional problem to manage: the imports of sugar coming from various categories of countries to which it has provided preferential access. The EU sugar policy implements a whole range of instruments, including production quotas, intervention prices, minimum prices for sugar beet, tariff rate quotas, prohibitive over-quota tariffs, and export refunds.

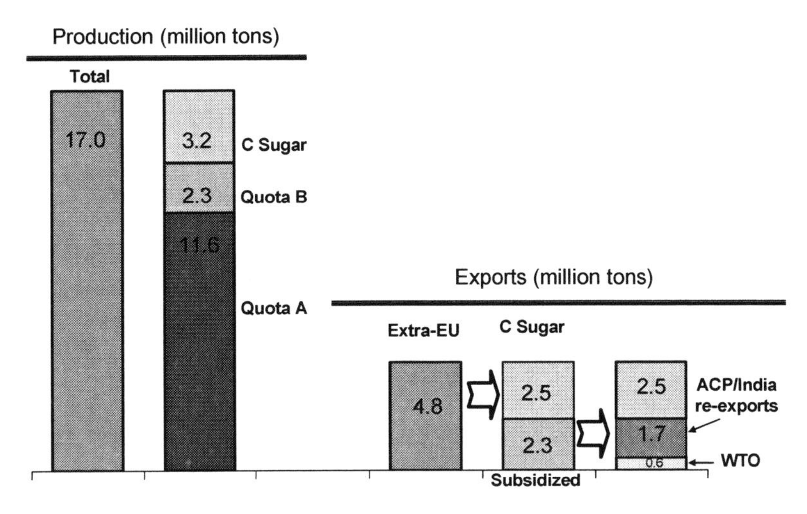

Source: EC/DGA

Figure 13. *Structure of production and exports of sugar in the EU-15 (2004/05)*

Figure 13 explains the situation under the EU sugar regime in 2004/05, before the reform of 2006. Domestic production equalled 17.0 million tons. Of this, 11.6 million tons fell under the A quota and was intended for domestic consumption. An intervention price of € 632 per ton established a price floor for this sugar. Another 2.3 million tons fell under the B quota. It was eligible for the intervention mechanism but subject to a levy of maximally 30 percent of the price. This levy was (and is) used to export 1.2 million tons of this B sugar (the maximum to which the EU has committed itself in the WTO) with an export refund. In addition, the EU had two other kinds of sugar exports. It was these that were challenged by Brazil, Australia and Thailand in the sugar panel. The first was the over-quota production, called C sugar, which was exported without an export refund. In 2004/05, the EU produced 3.2 million tons of C sugar, exporting 2.5 million tons and carrying over 0.6 million tons to the following crop year. The panel found that the subsidies granted to A and B sugar had spill-over effects on C sugar, which was in this sense also exported with export subsidies (Watkins 2002; 2004).

In the second place, the EU exported a volume equivalent to the 1.7 million tons of white-sugar equivalents that it imported free of tariffs from countries to which it provided preferential access. The EU has established five different types of such preferences: the ACP Protocol and India Agreement, with a tariff rate quota of 1.3 million tons; the OCT quota for EU overseas countries and territories; the tiny CXL quota (82,000 tons from Brazil and Cuba); the 'Balkans' initiative (free access for Albania, Bosnia-Herzegovina, Croatia, FYROM and Serbia and Montenegro); and the Everything-but-Arms initiative through which LDCs are to gain gradually free access[11]. The preferential imports from ACP countries are highly concentrated in a few suppliers. Of the 1.3 millions tons imported under the ACP protocol and the India Agreement, 68 percent came from Mauritius, the Fiji Islands, Jamaica and Swaziland, none of which are LDCs. On the other hand, important poor sugar-producing countries such as Ethiopia, Mozambique, Sudan, Zimbabwe and Malawi have no quotas or quotas that are very small compared to their export potential. The subsidized export by the EU of a volume of sugar equivalent to that imported under these various preferential agreements was the other modality that was challenged by Brazil, Australia and Thailand, and condemned by the sugar panel.

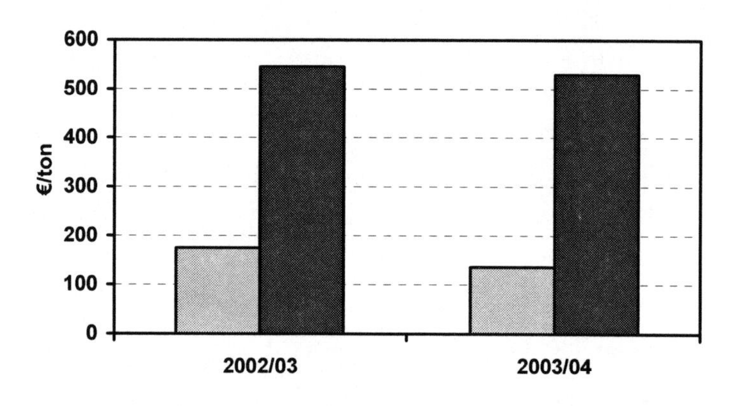

Source: EC/DGA; NY Board of Trade

Figure 14. *World sugar prices and EU import values from ACP countries*

As a response to the panel's verdict, the EU approved an effective reform of its sugar regime in 2006[12]. Intervention prices for white sugar and minimum prices for sugar beet were reduced and the difference between A and B sugar eliminated; temporary financial assistance was offered to beet and sugar producers in countries that would reduce their quota production by more than 50 percent; and domestic surpluses were to be managed by subsidies for private storage, intervention purchases, and stimulating of non-traditional uses like bio-ethanol. It is expected that sugar producers will opt for receiving the financial assistance as they will not be able to produce cost-effectively under the new domestic prices. As a consequence, production will be reduced even though the production quotas themselves remain unchanged.

Although the sugar reform does not result in significant changes in the provisions for imports, the reduction in domestic prices will decrease the prices received by preferential suppliers. This, in turn, will reduce distortions in the world market because the prices paid for preferential imports were much higher than the free world-market prices (Figure 14). Nevertheless, the costs of this preference erosion for least developed countries are burdensome.

LIBERALIZATION OF THE WORLD MARKETS FOR COTTON AND SUGAR: IMPACTS ON DEVELOPING COUNTRIES

The previous sections described the world markets for cotton and sugar and the developed country policies that have distorted these markets. This section will discuss the impacts that trade liberalization will have on developing countries. In doing so, the cases of cotton and sugar should be analysed separately. For cotton, all exporting developing countries will benefit from liberalization. For sugar, however, the outcome is more complex.

Developing countries that benefit from preferential exports will see their preferences being eroded and their export prices reduced, but competitive developing countries will expand their exports and receive higher prices (Wohlgenant 1999; Van der Mensbrugghe et al. 2003).

In both cases, it is essential to define the liberalization scenario precisely, given that it will be a combination of reductions in trade barriers, domestic support, and export subsidies. At the time of writing, there were two main drivers of liberalization: (i) the decisions of the cotton and sugar panels; and (ii) the Doha Round negotiations. (Table 1 summarizes the forces in favour of, or against, the liberalization of the cotton and sugar markets.) For cotton, the implementation of the panel's decisions would lead to a significant adjustment of U.S. policies. Additionally, a successful completion of the Doha Round would add a reduction in the EU subsidies and an increased access to the Chinese market. For sugar, the implementation of the panel's decisions means the elimination of EU export subsidies granted to the exports of ACP/India equivalent sugar and the implicit export subsidies for C sugar. In addition, a successful Doha Round would open, even if only partially, the protected domestic markets of the EU, the U.S. and Japan.

Liberalization in cotton

The main findings of the cotton panel were that:

(i) the direct payments (DP) were not green box measures and should be subject to limits and commitments of AMS;

(ii) the export credit programs (GSM 102, GSM 103 and SCGP) and the UMP (step-2 payments) were export subsidies and should be eliminated since they are inconsistent with U.S. export subsidy commitments;

(iii) step-2 payments, marketing loan programs, production flexibility payments, direct payments, marketing loss assistance, counter-cyclical payments, crop insurance payments and cottonseed payments granted domestic support on a commodity-specific basis;

(iv) marketing-loan programs, step-2 payments, marketing loss assistance and counter-cyclical payments caused significant price suppression, and their adverse effects in the world market should be eliminated.

***Table 1**. Forces working for or against the liberalization of cotton and sugar markets*

	For	Against
Cotton	– Cotton panel: (i) eliminate export payments (Step-2 and export credit programs); (i) eliminate adverse effects of domestic support payments (MLP and CCP) – Doha Round: reduce trade-distorting domestic support – Sectoral initiative required by African countries demanding trade liberalization – U.S. domestic conditions: claims from agricultural sectors that receive small amounts of subsidies compared to 'grains' (including cotton) sectors	– U.S. domestic conditions: (i) cotton farmers' lobby; (ii) negotiations on 2007 Farm Bill; (iii) dominance of the legislative over the executive in the formulation of farm programs. – Restricted capacity of the WTO Dispute Settlement Body (DSB) to enforce panel recommendations
Sugar	– Sugar panel: (i) eliminate subsidies provided to exports of 'ACP/India equivalent sugar'; (ii) eliminate cross-subsidization of C sugar exports – Doha Round: (i) phase out all export subsidies; (ii) increase market access through tariff reduction and TRQ expansion	– EU domestic conditions: difficulties for the European Commission to implement a reform that will result in effective elimination of subsidized exports – ACP countries: revenue loss due to reduction of EU intervention price – Doha Round: preference erosion due to tariff reduction

The U.S. can choose between some alternative ways of implementing the panel's decisions. From the point of view of developing countries, the most desirable one is close to liberalization of the market. In this case, export payments and distorting effects of price-based subsidies would be eliminated, even if other domestic subsidies would not be. Once price-based subsidies have been eliminated, U.S. farmers will respond to market signals rather than government payments in taking their production decisions. As a consequence, U.S. production will be adjusted to an extent depending on the level of world prices (Poonyth et al. 2004; Sumner 2003).

Analysts of the impacts of cotton trade liberalization agree that domestic and export subsidies are the main reason for the market distortions; that liberalization would mainly lead to adjustment in production in the U.S. and the EU; and that the removal of developed-country subsidies would result in higher world market prices[13]. Nevertheless, the question of access to the Chinese market cannot be neglected. Since 2002, China has become the world's largest importer of cotton. At present, its imports even exceed its tariff rate quota. A successful WTO round would stimulate more access to the Chinese market through over-quota tariff reduction and TRQ expansion. Nevertheless, elimination of trade distorting support by the US is needed to ensure that China's increased imports will benefit developing countries and improve world-market prices rather than leading to new increases in US production.

Liberalization in sugar

Where sugar liberalization is concerned not all developing countries would gain by trade liberalization (Elbehri et al. 2000). Certainly, least developed countries will face preference erosion and revenue losses due to the reform of the EU sugar regime. The balance between winning and losing developing countries depends on how sugar liberalization is approached. One approach, which is central in the sugar case against the EU, is to eliminate export subsidization. Another approach, which is central in the Doha Round negotiations, is the elimination or reduction of trade barriers.

The main findings of the sugar panel were that:

(i) the export refunds granted to the export of ACP/India equivalent sugar violated the EU export subsidy commitments;
(ii) the EU sugar regime led to the cross-subsidization of the export of C sugar, which likewise violated the export subsidy commitments of the EU;
(iii) the EU was recommended to bring its regulations into line with its obligations with respect to export subsidies.

The main implication of this recommendation is that the EU is constrained by the quantity (1,273,500 tons) and value (€499.1 million) of its current WTO commitment regarding subsidized sugar export. Any export beyond that level involves export subsidies. Of the 5.3 million tons of white sugar exported in 2003/04, 1.1 million tons were exported with regular export subsidies and 4.2 million tons with subsidies that should be eliminated. Given that the guaranteed sugar price in the EU is far above the international price, the EU can only manage its excess supply by reducing its domestic production. The phasing out of the EU's subsidized exports will benefit all sugar exporting countries, mainly the developing ones which account for almost 80 percent of world exports.

Reforming the EU sugar regime was a necessary condition for eliminating subsidized exports without changing the market access granted under trade preferences (Mitchell 2004). Reforming it through the reduction of guaranteed prices was an option that has transferred the burden of adjustment to domestic producers but also to preferential partners. Even though the latter have not lost market access, their export revenues have decreased. On the other hand, the reform will lead to a more balanced distribution of preferential trade shares across the EU's partners, thereby rewarding the most competitive sugar producers. In that sense, the reform goes in the right direction because it tackles two distortions: non-competitive subsidized exports and sugar imports under preferences from a few countries that are privileged compared to all ACP countries.

A more far-reaching liberalization of sugar trade might be achieved by the reduction of trade barriers as an outcome of the WTO Doha Round. Many studies have shown that reduction or elimination of trade barriers will lead to a rise in world prices and a reduction of price volatility. Different liberalization scenarios show that the developing countries will capture the benefits in terms of export volumes and welfare gains, and that the U.S., the EU and Japan will see an increase in their imports and a reduction in their production.

The distribution of gains among developing countries will depend on how far-reaching the liberalization will be. Multilateral liberalization will mainly benefit the most competitive developing countries. Partial liberalization, such as quota expansion on a non-MFN basis, will mostly benefit the countries with preferential access. A combination of this with over-quota tariff reduction can lead to a more balanced distribution of gains among countries with and without preferential access. A comparison of different scenarios by van der Mensbrugghe et al. (2003) shows that both under a partial and a full liberalization, most of the gains go to developing countries, and all developing countries will benefit. Even in the full-liberalization scenario, the gains will be fairly distributed across developing countries, although the most competitive among them will benefit more.

The main concern related to any liberalization scenario is the problem of preference erosion. However, even among developing countries with preferential access in the U.S. and the EU, this issue is controversial. Many competitive African and Central-American sugar producers wish to have increased access to U.S. and European markets, and are unhappy with the fact that the allocation of the import quotas over countries is constraining their export volumes. For these countries, preference erosion is not a major concern.

Conversely, preference erosion is a real problem for some developing countries that are highly dependent on U.S. and EU imports. Countries that can only export for high prices and face a prospect of preference should be compensated by the U.S. and the EU. The most competitive agricultural developing countries such as Mercosur members Chile, Mexico and Colombia should contribute by offering duty-free and quota-free access to least developed countries.

FINAL REMARKS

Both for cotton and sugar, developing countries are the largest producers and exporters and face unfair competition from subsidized exports and closed markets of many developed countries. In addition, developing countries are also the largest importers of these products, meaning that their markets are more open and their policies less trade-distorting than those of developed countries. Trade liberalization, therefore, first of all means the reform of developed countries' policies. This will benefit producers and exporters in developing countries. In the short run, they will benefit from a rise in prices and, in the long run, from new market opportunities that will be opened as developed countries withdraw subsidized exports and provide more access to their domestic markets.

The hard nut to crack in all this will be enforcing reform on developed countries' policies. The cotton and sugar cases and the agricultural negotiations in the Doha Round are the two vectors of change. Both will have a positive influence but the final outcome is still uncertain.

For cotton, both the panel's recommendations and the WTO negotiations may entail a reduction in the distorting effects of U.S. policies. Implementation of the panel's recommendations will involve the elimination of exporting payments through the reform of the export credit programs and the removal of step-2 payments. However, reducing the price suppression effects of price-based payments will depend on new farm policy legislation. The debates on the 2007 farm bill have already started in the U.S. and their outcome will be directly related to the reduction in domestic support on which parties will agree in the Doha Round. If a considerable reduction in domestic support is agreed upon, the U.S. administration will be obliged to negotiate a more far-reaching reform of the farm bill with the Congress. If the WTO negotiations lose ambition in this respect, the U.S. will be encouraged to keep its farm programs largely unchanged.

In the case of sugar, the WTO case has accelerated the reform of the EU sugar regime that the European Commission has already been trying to introduce since the early 1990s. Before the conclusion of the WTO case, the Commission mainly aimed at the elimination of the export refunds. The sugar panel's recommendation has obliged the EU to implement a broader reform because the cross-subsidization of C-sugar exports must also be ended.

By reforming its sugar regime, the EU will be prepared to implement a number of possible results of the Doha Round negotiations: (i) phasing out export subsidies; (ii) reducing import tariffs (at least the 'water in the tariffs') without exposing its domestic market to the international market; and (iii) reducing domestic support through the reduction of domestic prices. Besides, the reform will prepare the EU for the opening of its sugar market for LDCs in the frame of the Everything-but-Arms initiative.

Whether EBA countries will benefit from the reform depends on the level of domestic prices in the EU once the transitional period is completed. Assuming that these will follow the level of cost of the most competitive EU country (say, France), many EBA countries will still be in a position to supply sugar to the EU. Those that are less competitive will lose market share. The ACP countries among them should be compensated for the ensuing loss in export earnings.

The WTO negotiations can also cause the U.S. to reform its sugar policies. The tariff reduction and quota expansion mandated by the July 2004 'Doha Work Programme' will improve access to the American market.

Reform of trade-distorting agricultural policies of developed countries can only be achieved by multilateral negotiations and dispute settlement. Therefore, the WTO and the developing-countries' position in it should be strong. The formation of the G-20 was a very important step in this direction. At the time of writing, the results of the Doha Round were still unknown. However, it can be expected that the outcomes will be more in line with the needs of developing countries than those of the Uruguay Round.

NOTES

1 White-sugar weight is converted to raw-equivalent weight by a factor 1.08. All information about quantity mentioned in this paper is denominated in raw equivalent.

2 This figure does not take into account the intra-EU trade.

3 Saudi Arabia and the United Arab Emirates import raw sugar from Latin-American countries and sell the excess of white sugar to border countries.

4 The free market is the total trade minus the preferential imports and in-quota imports. In order to calculate the free market based on the USDA data, it is necessary to subtract from the total market the U.S. imports, the EU trade (exports and imports) – which already include the ACP and India exports to the EU – and the exports of Cuba to Russia. It is not necessary to perform the same calculation for the intra-tons EU trade because the USDA data do not take it into consideration.

5 The total amount of transfers under the MLP depends on the total current production and on the difference between the loan rate and the adjusted world price (the AWP is equivalent to the CIF Northern Europe price adjusted to U.S. base quality and average location). The marketing loan programs – Marketing Assistance Loan, Loan Deficiency Payment and Certificate Exchange Gains – although not working in the same way, have a very clear objective: to cover the price-gap differential between the market prices and the fixed rate (loan rate).

6 In order to follow similar principles of direct payments, payment acreage is set at 85% of base acreage.

7 This payment was activated when the following two conditions were met: (i) the price of U.S. cotton delivered in Northern Europe (USNE) exceeded the Northern Europe (NE) price by more than 1.25 cents per pound for four consecutive weeks; and (ii) the AWP was within 134% of the base loan rate. Payments were made available to domestic users of cotton consumed at the mill and to eligible exporters. The payment rate is equal to the difference in the fourth week of the four-week period between the USNE and the NE prices.

8 In order to respond to the cotton panel's findings, the U.S. government has eliminated the GSM-103 program and has implemented some modifications to GSM-102 and SCGP programs. Since October 2005, both programs work with risk-based fees as a way to ensure that such fees cover long-term operating costs and losses.

9 Four other trade-distorting instruments were notified in the 1995-2001 period: the Georgia Cotton Indemnity (1998); Cotton Seed Payments (1999 and 2000); Storage Payments (1997-2001); and the Commodity Loan Interest Subsidy (1998-2001).

10 Today many countries with quotas are not able to produce sugar and they must import to complete their volumes.

11 The ACP/India, CXL and EBA quotas apply only to raw sugar for refining. If refineries cannot source sufficient quantities via these quotas, a tariff quota at zero duty for raw cane sugar for refining, known as Special Preferential Sugar (SPS), is open to ACP countries.

12 Common organization of the markets in the sugar sector (COM), COUNCIL REGULATION (EU) n° 318/2006 of 20 February 2006.

13 See FAO (2004) for a full analysis of those studies.

REFERENCES

Baffes, J., 2003. *Cotton and Developing Countries: a case study in policy incoherence*. World Bank, Washington. Trade Note no.10. [http://siteresources.worldbank.org/INTRANETTRADE/Resources/TradeNote10.pdf]

Baffes, J., 2004. *The cotton problem*. World Bank, Washington. [http://www.fao.org/es/ESC/common/ecg/47647_en_CottonProblem_Baffes.pdf]

CIE, 2002. *Targets for OECD sugar market liberalisation: report prepared for the Global Alliance for Sugar Trade Reform and Liberalisation*. Center for International Economics, Canberra. [http://www.thecie.com.au/publications/CIE-OECD_sugar_market_liberalisation.pdf]

Dymock, P., 2002. *Working Party on Agricultural Policies and Markets: background information on selected policy issues in the sugar sector*. OECD, Paris. AGR/CA/APM(2001)32/FINAL. [http://www.olis.oecd.org/olis/2001doc.nsf/8d00615172fd2a63c125685d005300b5/b58856619f383e33c1256bd1004ceff4/$FILE/JT00127791.PDF]

Elbehri, A., Hertel, T., Ingco, M., et al., 2000. *Partial liberalization of the world sugar market: a general equilibrium analysis of tariff-rate quota regimes: paper presented as a selected paper at the annual conference on global economics analysis, June 28-30, 2000, Melbourne, Australia*. [http://www.monash.edu.au/policy/conf/10Hertel-update.pdf]

FAO, 2004. *Cotton: impact of support policies on developing countries: a guide to contemporary analysis*. FAO, Rome. FAO Trade Policy Technical Notes no. 1. [ftp://ftp.fao.org/docrep/fao/007/y5533e/y5533e01.pdf]

Gillson, I., Poulton, C., Balcombe, K., et al., 2004. *Understanding the impact of cotton subsidies on developing countries*. Overseas Development Institute, London. [http://www.odi.org.uk/iedg/Projects/cotton_report.pdf]

Goreux, L., 2004. *Prejudice caused by industrialised countries subsidies to cotton sectors in Western and Central Africa*. 2nd edn. [http://www.fao.org/es/esc/common/ecg/47647_en_Goreux_Prejudicef.pdf]

ICAC, 2002. *Production and trade policies affecting the cotton industry*. International Cotton Advisory Committee, Washington. [http://www.icac.org/meetings/cgtn_conf/documents/icac_ccgtn_report.pdf]

Karagiannis, G., 2004. *The EU cotton policy regime and the implications of the proposed changes for producer welfare*. FAO, Rome. FAO Commodity and Trade Policy Research Working Paper no. 9. [ftp://ftp.fao.org/docrep/fao/007/j2732e/j2732e00.pdf]

Mitchell, D., 2004. *Sugar policies: opportunity for change*. World Bank, Washington. World Bank Policy Research Working Paper no. 3222.

Poonyth, D., Sarris, A., Sharma, R., et al., 2004. *The impact of domestic and trade policies on the world cotton market*. FAO, Rome. FAO Commodity and Trade Policy Research Working Paper no. 8. [ftp://ftp.fao.org/docrep/fao/007/j2731e/j2731e00.pdf]

Sumner, D., 2003. *A quantitative simulation analysis of the impacts of U.S. cotton subsidies on cotton prices and quantities*. [http://www.mre.gov.br/portugues/ministerio/sitios_secretaria/cgc/analisequantitativa.pdf]

Van der Mensbrugghe, D., Beghin, J. C. and Mitchell, D., 2003. *Modeling tariff rate quotas in a global context: the case of sugar markets in OECD countries*. Iowa State Universty, Ames. Center for Agricultural and Rural Development Working Paper no. 03-WP 343. [http://www.card.iastate.edu/publications/DBS/PDFFiles/03wp343.pdf]

Watkins, K., 2002. *Cultivating poverty: the impact of US cotton subsidies on Africa*. Oxfam, Washington. Oxfam Briefing Paper no. 30. [http://www.oxfam.org.uk/what_we_do/issues/trade/downloads/bp30_cotton.pdf]

Watkins, K., 2004. *Dumping on the world: how EU sugar policies hurt poor countries*. Oxfam, Washington. Oxfam Briefing Paper no. 61. [http://www.oxfam.org.uk/what_we_do/issues/trade/downloads/bp61_sugar_dumping.pdf]

Wohlgenant, M.K., 1999. *Effects of trade liberalization on the world sugar market*. FAO, Rome. [http://www.fao.org/DOCREP/005/X2643E/X2643E00.HTM]

CHAPTER 7

HOW TO INCREASE THE BENEFITS OF THE DOHA DEVELOPMENT ROUND FOR THE LEAST DEVELOPED COUNTRIES

DAVID BLANDFORD

Professor of Agricultural Economics, Pennsylvania State University, University Park, Pennsylvania, USA

INTRODUCTION

Until the Uruguay Round in 1986-94, the interests of developing countries did not figure prominently in the series of trade negotiations undertaken under the General Agreement on Tariffs and Trade. The Uruguay Round Agreement contained provisions for special and differential treatment for developing countries. For example, the Agreement on Agriculture (AoA) provided an extended period for the implementation of agreed reductions in domestic support, export subsidies and tariffs for developing countries; the least developed countries (LDCs) were exempted from such reductions. Following the failure of the third WTO ministerial meeting in Seattle in 1999 the interests of developing countries were much more prominent at the meeting in Doha, Qatar in 2001. The Ministerial Declaration from that meeting that launched the current round of WTO negotiations contains no less than 24 references to developing countries, and 26 references to the least developed countries. The round has since come to be known as the Doha Development Round.

This paper assesses what can be done to increase the benefits for Least developed Countries (LDCs) from a new WTO agreement. To a large extent any assessment of the balance of advantages and disadvantages has to be conjectural at this stage; much of the detail remains to be determined. Only a framework for modalities was established at the negotiating session in Geneva in July 2004. Nevertheless, the content of that framework and its potential implications for LDCs are assessed. While much of the focus is on the implications of a new agreement for agriculture, because of the importance of that sector for LDCs, other areas of

N. Koning and P. Pinstrup-Andersen (eds.), Agricultural Trade Liberalization and the Least developed Countries, 105–128.

concern, such as tariffs on non-agricultural goods such as textiles and apparel, are noted.

THE TREATMENT OF DEVELOPING COUNTRIES IN THE GATT/WTO

Several developing countries were involved in the creation of the General Agreement on Tariffs and Trade (GATT) in 1947 – 12 of the original 23 contracting parties to the GATT would not have been classified as industrial countries at the time[1]. Roughly two thirds of the current 148 members of the WTO are identified as developing countries. The original treaty did not provide any special treatment for these countries. The fundamental principles of the GATT – non-discrimination through the application of the Most Favoured Nation (MFN) principle and equality of treatment with domestic products (national treatment) were supposed to apply to all signatories. However, in 1955 a revision of Article XVIII, dealing with government assistance to economic development, introduced some flexibility for those contracting parties "the economies of which can only support low standards of living and are in the early stages of development" in the use of quantitative restrictions to address balance of payments problems and in the use of tariffs to promote the development of a particular industry. Further changes were introduced in 1965 through Article XXXVI on trade and development. In that article the developed countries identify the reduction and elimination of barriers to trade for the products of developing countries as a high priority. The article also introduces the concept of non-reciprocity in trade negotiations between developed and developing countries, i.e., the extension of trade concessions by developed countries that are unmatched by concessions by developing countries. A subsequent decision in 1979 known as the 'Enabling Clause' solidified the concept of special and differential treatment and non-reciprocity in trade negotiations. It legitimized preferential tariff treatment for the exports of developing countries within the Generalized System of Preferences (GSP) and provided differential and more favourable treatment on provisions relating to non-tariff measures; it sanctioned regional or global arrangements for the reduction or elimination of tariffs among developing countries; and provided for special treatment for the least developed countries in the context of measures for developing countries as a whole[2]. Of the 50 countries currently identified by the United Nations as least developed countries, 32 are members of the WTO (Table 1), a further 8 are in the process of accession and 2 are WTO observers.

The Uruguay Round Agreement contains special measures that recognize the interests of developing countries. These relate to: 1. provisions that address the interests of developing and least developed countries in a general manner; 2. an easing of the rules or obligations to be met under the Agreement; 3. the provision of a longer time-frame for the implementation of commitments; and 4. technical assistance. In the Agreement, LDCs were required to make fewer commitments than other countries and WTO members were encouraged to use a fast track approach for the application of concessions on tariffs and non-tariff measures for imports of particular relevance to LDCs.

***Table 1**. Selected agricultural trade characteristics of LDCs*

Country *	WTO status	Share of ag. in GDP	Ratio of ag. trade to GDP	Leading ag. export	Share in total exports	Share of food imports in total imports
		percent	percent		percent	percent
		1998	1998	1997-99	1997-99	1997-99
Sao Tome	O	21	41	Cocoa beans	69	25
Malawi	M	36	36	Tobacco leaves	59	13
Kiribati	N	21	34	Copra	42	26
Mauritania	M	25	29	Cattle	4	78
Gambia	M	27	26	Groundnuts, shelled	20	37
Sierra Leone	M	44	23	na	na	86
Guinea-Bissau	M	62	23	Cashew nuts	48	22
Djibouti	M	4 [a]	22	Cattle	18	24
Lesotho	M	12	21	Wool	2	13
Yemen	A	18	19	Coffee, green	<1	30
Solomon Is.	M	na	19	Palm oil	10	11
Samoa	A	42 [b]	19	Copra	12	20
Vanuatu	A	25 [a]	18	Copra	43	12
Maldives	N	16	16	na	na	10
Mali	M	47	16	Cotton lint	30	10
Togo	M	42	15	Cotton lint	23	14
Cape Verde	A	12	15	Apples	1	24
Benin	M	39	14	Cotton lint	33	16
Niger	M	41	13	Cigarettes	9	24
Senegal	M	17	13	Groundnut oil	3	30
Burkina Faso	M	33	12	Cotton lint	39	14
Comoros	N	39	11	Vanilla	34	40
Ethiopia	A	50	11	Coffee, green	62	10
Burundi	M	54	10	Coffee, green	22	12
Chad	M	40	10	Cotton lint	37	9
Eritrea	N	17	10	Sesame seed	4	11
Bhutan	A	38	9	Oranges	4	9
Tanzania	M	46	9	Cashew nuts	16	16
Haiti	M	30	9	Coffee, green	8	35
Uganda	M	45	9	Coffee, green	54	11

Table 1 (cont)

Table 1 (cont)

Country *	WTO status	Share of ag. in GDP	Ratio of ag. trade to GDP	Leading ag. export	Share in total exports	Share of food imports in total imports
		percent	percent		percent	percent
Zambia	M	17	9	Sugar	2	17
Sudan	A	39	8	Sesame seed	13	12
Centr. Afr. Rep.	M	53	7	Cotton lint	11	15
Rwanda	M	47	7	Coffee, green	43	30
Angola	M	12	6	Coffee, green	<1	15
Eq. Guinea	O	22	6	Cocoa beans	2	23
Cambodia	M	51	5	Rubber	9	27
Madagascar	M	31	5	Coffee, green	12	15
Laos	A	53	4	Coffee, green	4	3
Congo, D.R.	M	58 a	4	Coffee, green	10	38
Nepal	M	41	4	Wheat flour	5	7
Bangladesh	M	22	3	Jute	2	18
Afghanistan	N	na	na	Skins (goats)	14	18
Liberia	N	na	na	Rubber	9	14
Myanmar	M	53	na	Beans, dry	13	7
Somalia	N	65	na	Cotton lint	23	50
Timor	N	na	na	na	na	na
Tuvalu	N	na	na	na	na	14

* Ordered on the basis of the ratio of agricultural trade to GDP (where data exist).

M = member; A = in process of accession; O = observer; N = non-member

a = 1997; b = 1993; na = not available

Sources: WTO website and FAO (2002).

The AoA, which represented the first serious attempt to liberalize agricultural trade within the framework of the GATT, contains provisions on market access, export subsidies and domestic support. Bound tariffs were established and reduced by an agreed percentage; imports of some products at lower rates of duty were managed though tariff-rate quotas (TRQs). The permitted value of export subsidies and the volume of subsidized exports were capped and reduced. Limitations were placed on the amount of trade-distorting (Amber Box) domestic support through the use of the concept of the aggregate measure of support (AMS) and the maximum permissible amount of that support was reduced. For developing countries, the required reduction in tariffs and in Amber Box support was lower than for developed countries, and the implementation period was longer (10 years rather than 6 years). Developing countries were granted a higher level for Amber Box support that was exempted from the AMS reduction commitment – the so-called *de minimis* level. This was set at 10% of the relevant value of production as opposed to 5% for developed countries (i.e., the value of production of an individual commodity for the commodity-specific *de minimis* and value of total production for the non-product-specific *de minimis*). Certain forms of domestic support that are part of development programs were also exempted from the reduction requirement. Most important, the LDCs were not required to make any commitments on market access, domestic support or export subsidies.

The Uruguay Round resulted in several other important agreements. From the perspective of the export interests of LDCs one of the more significant was the Agreement on Textiles and Clothing (ATC). Prior to that agreement, a substantial share of world trade in textiles and clothing was regulated by import quotas. Under the ATC quotas were gradually relaxed until their final elimination on January 1, 2005.

THE CURRENT TRADE LIBERALIZATION PROPOSALS

In launching the current round of international trade negotiations at the Doha meeting in November 2001, the WTO ministers declared "we shall continue to make positive efforts designed to ensure that developing countries, and especially the least developed among them, secure a share in the growth of world trade commensurate with the needs of their economic development. In this context, enhanced market access, balanced rules, and well targeted, sustainably financed technical assistance and capacity-building programmes have important roles to play." (WTO 2001, paragraph 2). The declaration also states "We recognize the particular vulnerability of the least developed countries and the special structural difficulties they face in the global economy. We are committed to addressing the marginalization of least developed countries in international trade and to improving their effective participation in the multilateral trading system." (WTO 2001, paragraph 3).

Since the Doha meeting, it has proved difficult to reach agreement on the details of a package of trade reform measures, particularly for agricultural products. Following the failure of the following Ministerial meeting in Cancún in September 2003, a Framework Agreement for completing the negotiations was finally concluded in August 2004. According to the Ministerial declaration this agreement is intended to provide "the additional precision required at this stage of the negotiations and thus the basis for the negotiations of full modalities in the next phase" (WTO 2004, paragraph A-1).

The framework agreement on agriculture

The major elements of the Framework Agreement for agriculture address the three principal elements (pillars) of the Uruguay Round AoA: domestic support, export competition and market access. The principal components are:

Domestic support

- Substantial reductions in overall trade-distorting support (defined as the sum of the total AMS, *de minimis*, and Blue Box support[3]) with a strong element of harmonization to be applied by developed countries – higher levels of support will be subject to deeper cuts through a tiered approach; product-specific aggregate measure of support (AMS) will be capped at average levels to be agreed.
- Reductions in the final bound total AMS and *de minimis* levels, a capping of payments at 5% of the total value of production with respect to an agreed historical period, and a capping of the AMS for individual commodities.
- The criteria for Green Box support (identified as minimally production and trade distorting) are to be reviewed and clarified to ensure that payments have no, or minimal, trade distorting or production effects; there is to be improved monitoring and surveillance of such payments.

Export competition

- Elimination of export subsidies.
- Elimination of export credits, credit guarantees or insurance programs with repayment periods beyond 180 days; disciplines to be imposed on shorter-term credits.
- Elimination of trade-distorting practices of exporting state trading entities (STEs).
- Disciplines to be imposed on certain types of food aid with the aim of preventing the displacement of commercial sales.
- These measures to be implemented in a phased manner by an agreed end date.

Market access

- Reductions in tariffs from bound rates using a tiered formula that will produce deeper cuts in higher tariffs.
- A 'substantial improvement' in market access for each product to be achieved through combinations of MFN commitments on TRQs (increased quota levels) and tariff reductions.
- Members may designate some products as 'sensitive products' with a given number of tariff lines (to be negotiated) that will be subject to less liberalization.

The Framework Agreement states that "special and differential treatment for developing countries will be an integral part of all elements of the negotiation, including the tariff reduction formula, the number and treatment of sensitive products, expansion of tariff rate quotas, and implementation period" (WTO 2004, paragraph 39). Specific provisions for developing countries are:

- Domestic support – longer implementation periods and lower reduction coefficients for trade distorting domestic support. Exemption from reductions for countries that allocate 'almost all' *de minimis* support to subsistence and resource-poor farmers.

- Export competition – longer implementation periods for the gradual elimination of all forms of export subsidies and differential treatment for least developed and net food-importing countries with respect to disciplines on export credits, guarantees and insurance programs. Ad hoc financing arrangement may be agreed in exceptional circumstances to meet import needs. STEs in developing countries that preserve domestic price stability and food security are to receive special consideration in retaining their monopoly status.
- Market access – smaller tariff reductions or tariff quota expansion commitments than for developed countries. Flexible treatment for products designated as 'special products'. Creation of a special safeguard (SSG) mechanism for developing countries to address surges in imports. Full implementation of the commitment to liberalize trade in tropical products. The issue of tariff preference erosion 'will be addressed'.

The LDCs will have access to all the provisions applicable to other developing countries. In addition, they will not be required to make any reduction commitments.

Cotton

Cotton is an important commodity for a number of LDCs; cotton policies in developed countries proved to be a contentious issue at the Cancún ministerial meeting. In the run up to the meeting Benin, Burkina Faso, Chad and Mali launched a joint 'cotton initiative' to address the impact of subsidies provided to cotton producers in developed countries. The WTO held a workshop on cotton in Benin in March 2004 to address the development assistance aspects. As a result of these efforts, the Doha work program document includes a specific section on cotton that "reaffirms the importance of the Sectoral Initative on Cotton" (WTO 2004, paragraph 1). That section indicates that the trade-related aspects of the Initiative will be addressed in the agricultural negotiations. It should also be noted that Brazil brought a successful case against U.S. cotton policies under the WTO dispute settlement procedure in 2004. As a result of the judgment in that case, the United States may make changes in a range of measures that were judged to have depressed world cotton prices.

Other provisions

Developing countries will be affected by other elements of a final agreement; perhaps the most significant will be the final package of tariff reductions for non-agricultural products. The framework for market access for such products involves the application of a non-linear formula for the reduction or elimination of tariff peaks, high tariffs, and tariff escalation. These issues are of particular relevance to some commodities of particular relevance to LDCs, such as textiles (see below). There will be an attempt to increase the proportion of tariffs that are bound, to convert specific tariffs to bound *ad valorem* equivalents, and to eliminate low tariffs. As for agriculture, developing countries will be given greater flexibility in making tariff cuts and will have a longer implementation period.

LDCs will not be required to apply the agreed formula for tariff reductions but will be expected to increase the proportion of their tariffs that are bound. In addition, the agreement calls upon "developed country participants and other participants who so decide, to grant on an autonomous basis duty-free and quota-free access for non-agricultural products originating from least developed countries" by a year to be determined (WTO 2004, Annex B, paragraph 10).

Other aspects of importance to developing countries, such as capacity constraints, the problems faced by small, vulnerable economies and the need for technical assistance, are mentioned. The particular interests of LDCs are noted specifically with respect to trade in services and trade facilitation.

POTENTIAL IMPLICATIONS OF THE PROPOSALS FOR LDCS

It seems clear from the WTO framework document that LDCs will not be asked to make any significant concessions with respect to tariffs on non-agricultural products or tariffs and other measures applied to agricultural products in the current round of negotiations. There appears to be a willingness to continue to expand the trade-related technical assistance provided to developing countries, particularly for the LDCs[4].

Trade theorists generally point to the global benefits that can be realized from trade liberalization, resulting from increased consumer choice and enhanced production efficiency through specialization. Countries that participate in trade negotiations focus more narrowly on the potential implications of freer trade for their balance of trade. In the light of this, the most immediate concern for the LDCs would seem to be the potential impact of an agreement on the competitive position of their exports, and how an agreement would affect the prices of their imports.

The reduction of tariffs and other trade barriers, coupled with reductions in trade-distorting support and export subsidies, would be expected to reduce distortions in domestic and international markets. From a competitive exporter's point of view, export prices would be expected to rise; export earnings would be expected to increase as a result of those higher prices and possibly through higher export volumes. Whether the world price effects of liberalization persist over time depends on the overall balance between global supply and demand. Whether any volume effect persists for an individual exporter depends on the long-run competitiveness of that exporter in international markets[5]. On the other side of the trade balance, an increase in world prices will affect import costs. Whether liberalization will improve or worsen the balance of trade cannot be determined *a priori*.

Agricultural products are not the principal source of export earnings for LDCs as a group. As may be seen from Figure 1, the value of agricultural exports ranked fourth among the leading commodity export groups in 2000-2001. Exports of textiles, for example, were more than 3 times larger than exports of agricultural products. However, much of the emphasis on the impact of further trade liberalization has been on agricultural products, because agriculture is a major sector of the economy in many LDCs and because domestic subsidies and trade-policy interventions are highly important for global agricultural trade. The OECD secretariat's estimate of total transfers to the agricultural sector from consumers and taxpayers in OECD member countries (around $950 million per day in 2003) is often used as an indicator for the magnitude of distortions[6]. Alternatively, the average bound tariff of over 45% for imports of agricultural products in industrialized countries, compared to an average tariff of around 4% for industrial products, might also be cited[7].

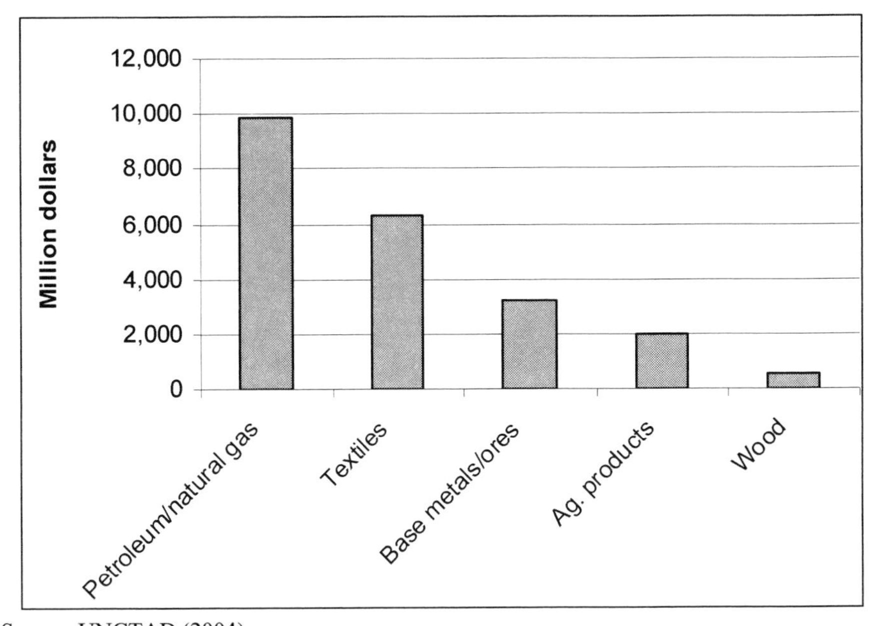

Source: UNCTAD (2004).

Figure 1*. Leading exports of LDCs in 2000-2001*

A number of studies have been conducted on the potential effects of further global trade liberalization. A recent study by Anderson et al. (2005) estimates the effects on trade volumes and real income of the complete liberalization of global merchandise trade by 2015. These show an increase in the volume of trade of 20% for all developing countries and an increase in real incomes of developing countries of 0.8%. The study does not present results for LDCs as a group, but shows increases in both trade volumes (23%) and real income (1.1%) for Sub-Saharan Africa that are higher than the developing country average. The authors note, however, that some LDCs are slight losers in simulations of partial liberalization when LDCs do not reduce their own trade barriers, due to the impact of a reduction in preference margins in developed countries. In a second study, Anderson and Martin (2005) suggest that the real incomes of low-income developing countries as a group would be roughly $16 billion higher with complete liberalization, even though their terms of trade (ratio of export to import prices) would decline. In both of these studies, a range of partial trade liberalization scenarios are shown to result in increases in real incomes for the low-income developing countries.

Studies conducted for agricultural products suggest that the elimination of distortions created by tariffs and subsidies would lead to higher world prices. Some relatively conservative estimates are provided by Diao et al. (2001), who indicate that agricultural commodity prices would increase by roughly 12% on average as a result of the elimination of trade distortions (Table 2)[8]. The price effects are greatest for commodities that are most heavily protected in developed countries, such as livestock products, wheat and other grains, sugar, oilseeds and rice. Developing countries that are net importers of food would be negatively affected by the increase in prices. On the other hand, some of these commodities are major exports for LDCs (Table 1). Many other commodities of importance to LDCs, such as tropical beverages, already face low tariff barriers in developed countries and would be little affected by liberalization.

***Table 2**. Effects of trade liberalization on world agricultural prices*

Commodity	Full liberalization	Removal of:		
		tariffs	domestic subsidies	export subsidies
	percentage change from base:			
Wheat	18	3	12	2
Rice	10	6	2	2
Other grains	15	1	12	1
Fruit and vegetables	8	5	0	3
Oilseeds and oil	11	3	8	0
Sugar	16	11	2	3
Other crops	6	4	1	0
Livestock products	22	12	6	3
Processed food	8	5	2	1
All products	12	6	4	2

Note: the sum of the figures for the individual sources of distortion does not necessarily equal the full liberalization percentage due to interaction effects.

Source: Diao et al. (2001).

The agricultural trade characteristics of LDCs can be seen from Table 1. Countries are ordered on the basis of the ratio of agricultural exports to GDP. For some, the necessary data are unavailable so those countries are listed alphabetically at the foot of the table. For the 44 countries for which data are available, in just over half (26) agricultural exports were equivalent to 10% or more of GDP; for roughly 20% of the countries (9) the ratio was over 20%. Data are available for 46 countries on the principal agricultural export commodity. Beverages (cocoa and coffee) are the leading export in 13 countries, cotton in 8 countries and oilseeds in 8 countries.

From Table 3 it may be seen that average tariffs are generally low for these commodity groups (beverages and tobacco, fibres, oilseeds) in the major developed countries. However, other products of interest to LDCs, such as sugar, meat and meat products, and to some extent fruit and vegetables face higher average tariffs. There might be the potential for increased export earnings for LDCs if such tariffs were reduced, although middle-income developing exporters (for example, Brazil and Thailand) might have the most to gain from a general reduction in tariffs in developed countries.

Table 3*. Agricultural tariffs by major commodity group in developed countries (%)*

Commodity	EU25	US	Developed Asia	EFTA	Developed Cairns
Paddy rice	5	6	288	18	0
Processed rice	4	4	287	11	0
Coarse grains	3	1	79	70	8
Wheat	11	3	106	136	6
Sugar	129	7	122	40	4
Oilseeds	0	3	77	46	0
Live animals	43	0	31	103	0
Animal products	8	1	11	48	12
Meat	98	6	23	197	7
Meat products	26	4	29	167	32
Dairy products	41	15	22	87	133
Fibres	0	9	0	0	0
Fruit and vegetables	10	3	21	37	2
Other crops	2	8	5	24	2
Fats	4	4	5	45	3
Beverages and tobacco	14	3	13	15	6
Food	11	5	12	23	9
Total agri-food	17	5	25	52	18

Source: Bureau et al. (2005).

While the emphasis in this chapter is primarily on agriculture, it should be stressed that trade liberalization in other sectors could be important for LDCs. Table 4 contains average applied tariff rates in selected countries for various categories of products. From the table the relatively high rates applied to agricultural products are evident, but it is also clear that manufactures, in particular textiles, face high average tariffs in some regions. The figures also suggest that there is significant tariff escalation (increase in the size of tariffs with the level of processing) for agricultural products in Western Europe and Japan, as well as in Latin America and South Asia. Figure 2 illustrates that tariff escalation is also an issue for textiles in many regions. Indeed such escalation is more pronounced in developing countries than in developed countries. Even though import quotas on textiles were eliminated on January 1, 2005, high tariffs are still applied to textile imports in many countries.

Finally, it may be noted that tariff peaks (defined as tariffs greater than 15%) are more prevalent for trade in manufactures among developing countries, than for trade between developing and developed countries (Figure 3).

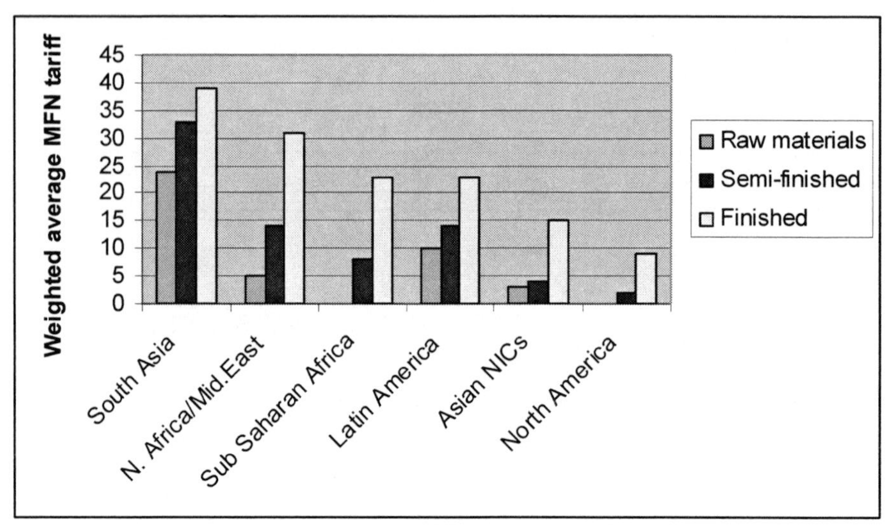

Source: UNCTAD (2003a).

Figure 2. *Tariff escalation for textiles*

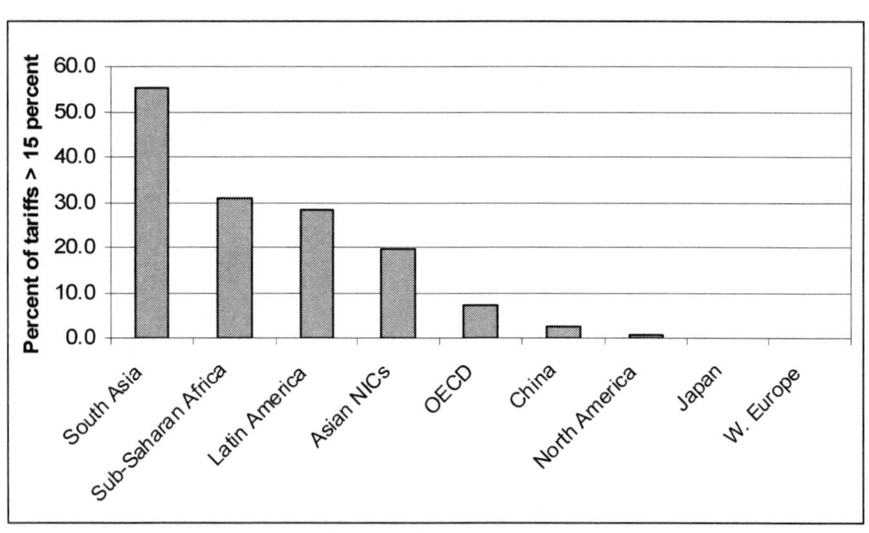

Source: UNCTAD (2003a).

Figure 3. *Tariff peaks on imports of manufactures from developing countries*

Table 4. *Average applied tariffs in selected importing countries and regions*

	Product group	China	South Asia	Latin America	Western Europe	North America	Japan
Natural resources	2	2	14	5	0	0	0
Primary agriculture	38	16	21	2	12	9	30
Processed agriculture	20	15	29	17	21	10	46
Textiles and apparel	8	13	28	15	5	10	6
Manufactures	5	6	24	11	2	1	0
Services	0	0	0	1	0	0	0

Source: UNCTAD (2003a)

The role of preferences

It might be concluded that given the prevalence of tariff barriers for agricultural products and labour-intensive manufactures, further trade liberalization would be advantageous for LDCs. Such a conclusion would need to be qualified by the fact that many LDCs already have preferential access for some of these products in developed countries. Preferential access is either provided through a reduction or elimination of the tariffs applied to LDC imports. In some cases, such concessions only apply to limited quantities of imports. The creation of the Generalized System of Preferences (noted earlier) has resulted in a number of preferential schemes. In addition, the European Union and the United States have regional schemes that benefit some LDCs. Box 1 summarizes the principal preferential schemes of relevance to LDCs in the four Quad countries (Canada, the European Union, Japan and the United States).

Relatively little empirical research has been conducted on the impact of tariff preferences on eligible countries (Tangermann 2002). It is difficult to estimate the short-term welfare gains resulting from preferences, and even more difficult to calculate any longer-term gains, for example, through the impact of preferential access on investment. Export earnings may increase due to the improved competitive position created for preference receiving countries relative to their non-preferential competitors. To the extent that preferential exporters are able to retain part of the preference margin, i.e., the difference between their supply price and the higher market price in the preference-granting country, this will also increase earnings. The opportunities for retaining such a preference 'rent' are greatest when preferential access is associated with a country-specific quota; otherwise much of the margin is likely to be captured by importing firms as suppliers compete for market share in the preference-granting country.

Some analysis has been undertaken of the effects of changes in tariff preferences, particularly in terms of the erosion of existing preferences that would be implied by a general reduction of MFN tariffs through the current round of trade negotiations.

Box 1: *Preferential tariff schemes for LDCs in the Quad countries.*

Canada

Market Access Initiative (MAI)
Introduced on January 1, 2003. All imports originating in an LDC (48 eligible countries) are granted duty-free, quota-free access with the exception of dairy, poultry and egg products which are subject to duties and quotas. Most of the 882 products affected by the Initiative are apparel and textile goods (760). A further 64 are food products and 43 are footwear items. LDCs were allowed duty-free access on a more restricted set of products since 1983 under Canada's General Preferential Tariff system – its GSP scheme. Cumulation of imports among countries eligible for GPT or the MAI is permitted.

European Union

Cotonou-Lomé (ACP Agreements)
A series of four Lomé agreements between 1975 and 2000 provided preferential access to the EU market for certain exports from African, Caribbean and Pacific (ACP) countries – the former colonies of the EU member states (includes 40 LDCs). Many agricultural products, particularly those supported under the EU's Common Agricultural Policy, were excluded. Some products (bananas, beef, horticultural products, rice, sugar, tobacco) were subject to low or zero tariffs up to a given level of imports. In 2000 the Cotonou agreement was signed. This will replace the non-reciprocal tariff preferences by a series of reciprocal Economic Partnership Agreements (EPAs) after 2007.

Everything But Arms (EBA) Initiative
From March 1, 2001 the European Union amended its existing GSP scheme to provide duty-free access for exports of all products (excluding arms) from 48 LDCs. Bananas, rice and sugar are subject to transitional arrangements and full liberalization will not occur until after 2009. Cumulation of imports among LDCs is not permitted under the EBA, but is allowed under the ACP agreements.

Japan

Generalized System of Preferences (GSP)
LDCs have duty and quota free access on a range of products under the GSP scheme. Other countries eligible for GSP have more restricted access and pay reduced tariffs. However, the list of eligible agricultural products for LDCs is limited. Some cumulation of imports among eligible countries is permitted.

Box 1 (cont.)

Box 1 (cont.)

United States

Generalized System of Preferences (GSP)
Eligible developing countries have duty-free access on roughly 3,000 products. LDCs are eligible for duty-free access on a broader range of products. The determination of eligibility is subject to a number of political and economic criteria. Currently 44 LDCs are eligible for duty-free access under general provisions of the GSP scheme and 41 are identified as LDC beneficiary countries. Products must meet a minimum value added requirement in an eligible country. Cumulation is allowed among eligible members of recognized associations, such as the Southern African Development Community (SADC). Duty-free access for a country may be subject to a quantitative limit if that country is determined to be too competitive.

African Growth and Opportunity Act (AGOA)
Signed in May 2000, the Act provides duty-free access on virtually all products in the GSP program for eligible countries in Sub-Saharan Africa. A more stringent set of criteria are applied than under the GSP to determine eligibility. As a result only 37 of the 45 African countries eligible under the GSP are also eligible under AGOA; 23 African LDCs are eligible. There are no quantitative limits on imports under AGOA.

Caribbean Basin Initiative (CBI)
The Initiative was introduced on January 1, 1984. Haiti benefits from duty-free access for its exports under this initiative. Product eligibility is similar to that under the GSP and AGOA.

Typically the methods used rely on fairly aggregate data and do not generate estimates for the LDCs as a group. In this chapter, the focus is on the impact of trade liberalization in general, rather than in particular commodity sectors such as agriculture. More detailed analysis of agricultural issues is contained in the chapter by Yu.

In a recent study, Alexandraki (2005) evaluates the impact of preference erosion for the G-90 countries, which includes both the LDCs and the ACP countries. She concludes that the impact of likely preference erosion under a new WTO agreement will be limited and that most of the effects will be confined to middle-income developing countries (Mauritius, St. Lucia, Belize, St. Kitts and Nevis, Guyana and Fiji) because of the implications for sugar and bananas and, to a much lesser extent, textiles.

In an earlier study, Subramanian (2004) examined the implications for LDCs of a 40% reduction in MFN tariffs for agricultural and manufactured goods in the Quad countries. The estimates were based on optimistic assumptions about the current gains accruing to LDCs from preferences, in particular, that the rules applying to preferential trade do not have any restrictive effects on their exports to the Quad countries. Under these assumptions Subramanian estimates that preference erosion would result in a reduction in the value of total LDC exports of less than 2%. Five LDCs face losses in excess of 5% of the value of their exports – Malawi (12%), Mauritania (9%), Haiti (6%), Cape Verde (6%) and Sao Tome and Principe (5%). In absolute terms, the larges losers are Bangladesh (US$ 222 million), Cambodia (US$

54 million); Malawi ($US 49 million), Mauritania (US$ 40 million) and Tanzania (US$ 29 million). On the basis of these estimates Malawi and Mauritania could face significant losses in both absolute and relative terms.

These studies focus exclusively on the erosion of existing preferences. As noted in Table 1, many existing schemes do not provide completely free access for imports from the LDCs. Some studies have analysed the expansion of preferences through the general application of a scheme similar to the Everything But Arms (EBA) initiative of the European Union to all imports from LDCs in the Quad countries. This would imply that the margin of preference provided to LDCs would be increased through the elimination of any remaining tariffs and the removal of any limitations on the volume of imports. The results of these studies are summarized by Achterbosch et al. (2003). They suggest that a strengthening of preferences in the Quad countries would increase the export potential of LDCs by 3-13%, primarily through the impact on textiles and clothing in Canada and the United States, and agricultural products in Japan. Hoekman et al. (2002) note that tariff peaks in the Quad countries have a disproportionate effect on LDC exports since such peaks tend to be concentrated in agricultural products (sugar, cereals and meat) and in labour-intensive products such as apparel and footwear. If the export potential that would be created by the strengthening of preferences were to be exploited by the LDCs, this would increase their economic welfare by 1-2%. Unfortunately, there seems to have been little analysis of the impact of extending preferences for LDCs to a broader range of importing countries, but it is likely that such an expansion could also help to increase the export potential of LDCs. The potential importance of this issue is discussed in more detail later in the chapter.

In conclusion, reductions in applied MFN tariffs resulting from a Doha Round agreement could result in the erosion of existing LDCs preferences. To some extent, this erosion could be offset by a further strengthening of those preferences in developed countries, by removing remaining restrictions on import volumes, eliminating any remaining tariffs, and ensuring that duty-free and quota-free access is extended to all products. For the few LDCs that may experience significant reductions in the value of their exports as a result of preference erosion, compensation could be provided through existing international financial mechanisms (Subramanian 2004).

ARE PREFERENTIAL SCHEMES IN THE LONG-TERM INTERESTS OF LDCS?

Proponents of free trade argue that partial trade liberalization is inferior to the complete elimination of barriers to trade. Neo-classical trade theory suggests that the elimination of trade barriers would maximize global welfare by enabling the world's resources to be used most productively. We should note that even in a free-trade world, the distribution of the increase in economic welfare within and among countries may be highly uneven since that is crucially dependent on the distribution of factors of production and the returns to those factors. Since the complete elimination of trade barriers seems to be a distant possibility, our attention must be directed to the merits of 'second best' or more limited approaches, such as that reflected by preferential schemes.

The arguments for preferences rest on the stimulus that these are assumed to provide to the economies of poorer countries, by increasing the demand for their exports in richer countries. By offering duty-free access for the exports of developing countries, their export industries are expected to expand, generating higher domestic income and employment. As their industries grow they may be able to increase their efficiency and exploit economies of scale, making them more competitive and enabling them to compete in non-preferential markets. This argument is crucially dependent on the existence of sufficient productive and export capacity in LDCs to take advantage of the economic incentives that are created by duty-free access (Wainio et al. 2005).

The arguments against preferences are that these can serve to lock the economies of preference-receiving countries into particular patterns of production that are not sustainable in the longer run, and create dependence on preference-granting markets. When a country imposes import restrictions, the relative prices of the affected products will increase. A country with preferential access will respond to the distorted prices of the protected market – its domestic resources may be drawn into the production of protected products, in the same way that the resources of the protecting country are drawn into such products. If the long-run prospect is for the eventual elimination of protection, industries in the preference-receiving country may face a similar issue of long-run sustainability as those in the preference-granting country.

Issues of distortion and dependency can be intensified when preferences are granted on a limited range of products. Countries may grant preferences on those products that are currently exported by poorer countries, rather than products that they might be able to export. The continued dependence of many poorer countries on a limited range of primary commodities can be criticized on this basis. Alternatively, countries may be reluctant to extend preferential access to 'import-sensitive' products in which poorer countries have a comparative advantage. Many parts of agriculture and labour-intensive manufactures, such as textiles and footwear in developing countries, have been affected by this approach to preference schemes in the European Union and the United States (GAO 2001).

It is difficult to escape the conclusion that preference schemes confer risks in terms of long-run sustainability. Particularly in countries that are likely to be highly dependent on trade in comparison to the size of their domestic economies, the development of a trade structure that could be undermined by the gradual elimination of preferences poses substantial risks. Balanced against that risk is the extent to which the development of export industries in preference-receiving countries can provide a stimulus to the overall development of their economies. Given the considerable economic challenges facing the LDCs, and the likelihood that the playing field of international trade is unlikely to be levelled in the foreseeable future, it could be argued that any measures that can be taken by the international community to stimulate the growth of exports by LDCs merit serious consideration.

STRENGTHENING PREFERENTIAL ARRANGEMENTS

As noted above, regardless of arguments over their advantages or disadvantages agreements that grant preferential access to markets in developed countries are important for developing countries. In recent years, there have been attempts to use these to give extra advantages to LDCs. This is reflected in Canada's Market Access Initiative, the Everything But Arms initiative of the European Union, and the LDC components of the Japanese and US GSP schemes. However, there are a number of limitations associated with existing preferential schemes that need to be addressed.

Eligibility. Schemes differ in terms of which countries are eligible for preferential access. While several countries follow generally accepted conventions on countries eligible for preferential treatment, such as the list of LDCs compiled by the United Nations, this is not always the case. For example, the LDCs eligible for special provisions in the US GSP only include 41 of the 50 countries currently on the UN list. Preferential arrangements for regional groupings, such as those provided to the African, Caribbean and Pacific (ACP) countries by the European Union, have traditionally been more limiting in terms of eligible countries. Indeed the European Union sought a waiver from the WTO for the current agreement (the Cotonou agreement) in 2001 because of this. Two GATT panels had earlier concluded the preferences (tariff and non-tariff) for the ACP countries provided under the Lomé treaty were contrary to GATT obligations[9]. US preferential schemes have always provided for the exclusion of certain countries on political or other grounds. The most recent example of preferential access, the African Growth and Opportunity Act (AGOA) continues that tradition.

Product coverage. The amount of preferential access provided, in terms of product coverage and preferential rates of duty, differs significantly among agreements. The Japanese GSP scheme, even with the expanded product list for LDCs, provides only limited preferential access for agricultural products, but relatively broad access for industrial products. US preferential arrangements provide only limited access for textiles and footwear. The EBA scheme of the European Union seems to provide substantial potential access, once the transitional arrangements for sensitive products such as sugar are complete, by applying zero tariffs to imports from LDCs. Schemes that do not allow duty-free access can be designed to preserve a margin of preference for LDCs even with reductions in MFN tariff rates by expressing the preference as a percentage of the MFN tariff.

Rules of origin. There are substantial differences in rules of origin in preferential arrangements. Some involve the criterion of a change in tariff classification, others a percentage value added criterion and others apply criteria relating to manufacturing or processing. Some rules of origin discriminate against the integration of agricultural industries among developing countries and limit the opportunities for adding value to imported products that are subsequently re-exported. This is particularly problematic for small countries that would otherwise be able to develop a market for processed products, but are unable to provide their own raw materials. Several studies have suggested that the restrictiveness of rules of origin, the administrative burdens that these place on LDCs, and the resulting high transactions costs are responsible for low levels of utilization of preferences in some countries (e.g. Brenton 2003; Mattoo et al. 2002; UNCTAD 2003b). An Agreement on Rules of Origin was part of the Uruguay Round Agreement, but this was oriented towards the harmonization of non-preferential rules of origin. Apart from establishing some principles for the application of rules of origin under preferential agreements (essentially relating to transparency), the Agreement does not have anything to say about what types of rules are preferable. Member countries are merely required to notify the rules of origin that they apply under preferential agreements to the WTO.

Certainty of commitments. The amount of certainty on future market access provided under preferential programs differs considerably both in terms of the length of time to which agreements apply and whether countries can lose their eligibility. Schemes differ in the length of time for which they are in force. The Japanese GSP, for example, is renewed for significant periods of time. The current scheme extends to 2011. The renewal of the US scheme can be delayed by Congress. The scheme that expired in 1995, for example, was not renewed until 13 months later. Most schemes provide for the graduation of countries (loss of preferences) once a certain level of economic development is reached. That is relevant to LDCs if they are judged to have passed from the least developed to 'normal' developing country status. Of more immediate relevance is whether countries can lose their eligibility due to political factors. Most developed countries have suspended preferential access for Myanmar due to political conditions in that country, but many seem reluctant to take too active an approach to changing the list of eligible countries. The United States is an exception in this regard. For example, Chile and Paraguay were suspended from the GSP scheme during the 1980s on the basis of workers' rights before eventually being reinstated in 1991. Nicaragua's privileges were terminated in 1985 on the same grounds. After originally being included under AGOA, the Central African Republic and Eritrea were dropped as beneficiary countries from January 1, 2004 on the basis that they were not making sufficient progress towards policy reform[10]. The uncertainty created by the potential loss of eligibility for preferential access may reduce investment in export-oriented sectors in LDCs. Finally, countries that provide preferential access for developing countries still have the option of imposing higher tariffs on a 'temporary' basis, if the volume of those imports threatens to undermine prices in their domestic market. The existence of such 'safeguard' provisions adds an additional dimension of uncertainty to the commitments under preferential agreements[11].

Number of schemes. The number of countries that apply the generalized system of preferences for developing countries is limited. In addition to the GSP schemes operated by the Quad countries, a further 12 schemes have been notified to the UNCTAD secretariat. The countries involved are: Australia, Belarus, Bulgaria, the Czech Republic, Hungary, New Zealand, Norway, Poland, the Russian Federation, the Slovak Republic, Switzerland and Turkey[12]. It should be noted that some of these countries are classified as being in economic transition and some would not qualify as high-income countries on the basis of per capita GDP. It is noteworthy that under Part IV of the GATT, less developed contracting parties agree to take appropriate action in implementing the provisions of Part IV for the benefit of the trade of other less developed contracting parties. Furthermore, the framework agreement on agriculture for the Doha Round states that "Developed Members, *and developing country Members in a position to do so,* should provide duty-free and quota-free market access for products originating from least developed countries". (WTO 2004, paragraph 45, emphasis added).

Traditionally, the focus has been on the obligation of developed countries to open up their markets to developing countries. While this focus was undoubtedly justified in the past, one might question whether such a simple approach is appropriate for the future; particularly if one wishes to target preferences to LDCs. Currently, exports to other developing countries account for roughly 40% of LDCs total exports (UNCTAD 2004, Table 18). It is expected that income growth in developing countries will exceed that in developed countries for the foreseeable future. The IMF, for example, projected that average annual growth in real GDP for emerging market and developing countries will be roughly double that in advanced economies for 2004-09, with correspondingly higher growth in the volume of imports (Figure 4). Imports in developing Asia are projected to rise at an average rate of 14% per year. Furthermore, average applied tariffs are high and tariff peaks are common among developing countries, particularly for industrial products (UNCTAD 2003a). In order for the LDCs to take advantage of the opportunities for growth in exports to other developing countries, it would seem to be appropriate for many of the richer developing countries (e.g., Brazil, China, India, Malaysia) to develop preferential tariff schemes, targeted specifically at LDCs[13].

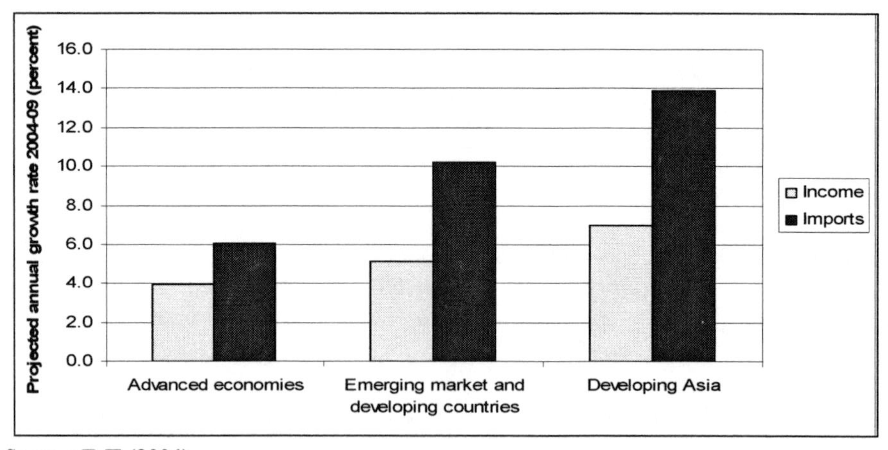

Source: IMF (2004)

Figure 4. *Projected growth of income and imports*

A Global System of Trade Preferences among developing countries (GSTP) was created in 1989 as a result of the efforts of the United Nations Conference on Trade and Development. There have been three rounds of negotiation under GSTP, the most recent in Brazil in 2004. However, only 44 developing countries have ratified the agreement, progress on negotiating improved market access has been slow, and the primary emphasis has been on reciprocal concessions rather than preferential access for LDCs. It should be noted that some developing countries have taken unilateral action to provide such access. Egypt, for example, has unilaterally reduced applied tariffs by 10-20% on 77 products of interest to LDCs and provided duty-free access for roughly 50 products.

Increasing the capacity of LDCs to take advantage of preferences. The application of preferences by a broader range of countries to all products, combined with improvements in how such preferences are implemented (as discussed above) could help to increase the opportunities for LDCs to broaden the range of products that they export and to increase their export earnings. However, it does not mean that LDCs would actually be able to profit from the new opportunities created. Many face considerable challenges in mobilizing their resources to take advantages to such opportunities. Some of these challenges are due to natural disadvantages created by geography, for example, distant location from expanding markets or land-locked locations. Some are characteristics of economic structures that constrain economic development, for example, limited natural resource or population bases. These factors can create substantial barriers to expanded participation in the global economy for some LDCs.

In other cases, barriers to export development can be overcome through a combination of appropriate domestic policies that stimulate investment in export-oriented industries, the development of the labour force and the acquisition and application of knowledge and information. Many factors are important in helping to overcome some of these barriers – including good governance, control of corruption, and the presence of a legal framework governing business transactions that is conducive to the development of a modern economy. Technical assistance, which has been increasingly emphasized by international organizations that work with LDCs, can also play an important role.

In the final analysis, expanded access to markets is a necessary but not sufficient condition for LDCs to be able to increase their participation in the global economy. It does not guarantee that this will be achieved.

CONCLUSIONS

It is unclear how much progress will be made in liberalizing international trade in the current WTO round of negotiations. Further reductions in tariffs and trade-distorting subsidies would likely lead to an increase in global economic welfare, but the impact on the least developed countries is unclear. The terms of trade of the LDCs could well deteriorate as a result of the impact of general reduction in MFN tariffs on the competitive advantage that LDCs secure through preferential access in some countries, and the price-enhancing effect that liberalization might exert on LDC imports. The effects on individual LDCs will depend on their export and import mix, and the extent to which the competitive position in their major markets is affected by a new agreement.

The negative effects of a general reduction in tariffs on the competitive position of the LDCs could be offset by strengthening their preferential access to markets in other countries. Regardless of the final outcome of the Doha Round, it seems likely that MFN tariffs in a range of products of importance to the LDCs will remain high, particularly in such commodity groups as agriculture and textiles. An expansion of duty-free/quota-free access to all LDC exports by developed countries could help to offset the erosion of tariff preferences, but this would not expand market access for LDCs in countries where the potential growth in demand is likely to be the strongest. To be really effective, the Doha Round would need to generate not only expanded access for LDC exports in developed countries, but also in higher-income developing countries. It is by no means clear that the Round will produce this result unless the LDCs (and their advocates) make a major effort to make this happen.

Furthermore, there are a number of significant limitations inherent in existing preference schemes for LDCs that need to be addressed. The schemes have a fragile legal status because they are created and managed on a unilateral basis by each preference-granting country. The terms of schemes vary considerably among countries. There is evidence that the rules of origin applied in existing schemes are burdensome for LDCs, result in limited trade creation, and actively discriminate against the development of value added exports. A second priority for LDCs in the current round of negotiations could be to bring preferential schemes under the GATT/WTO framework with the aim of establishing greater certainty and stability of access, provide unrestricted duty-free access for all LDC exports, and to simplify the rules to reduce transactions costs and actively encourage the development of value-added processing in LDCs.

If these measures were taken, they could help to achieve the stated objective of WTO ministers of improving the effective participation of LDCs in the multilateral trading system and satisfy their commitment to address the erosion of tariff preferences. But in the final analysis, such measures would only increase the potential for an expansion of exports by the LDCs. The least developed countries still face the challenge of mobilizing their economic resources to convert such potential into actual flows of goods.

NOTES

1. The developing countries were Brazil, Burma (Myanmar), Chile, China, Ceylon (Sri Lanka), Cuba, India, Lebanon, Pakistan, South Africa, Southern Rhodesia (Zimbabwe) and Syria. The other signatories were Australia, Belgium, Canada, the Czechoslovak Republic, France, Luxembourg, the Netherlands, the United Kingdom and the United States.
2. The concept of granting preferential tariff rates by developed countries for imports from developing countries was developed by Raúl Prebisch, the first Secretary-General of the United Nations Conference on Trade and Development (UNCTAD). It was formally adopted at the second UNCTAD conference in New Delhi in 1968. Certain preferential arrangements then in operation were exempted from MFN in the original GATT. In 1971 the Contracting Parties approved a waiver

to Article I in order to authorize the GSP scheme. The enabling clause created a permanent waiver for GSP.

[3] The Blue Box was included in the AoA to accommodate direct payments to producers made under production-limiting programs, primarily by the European Union and the United States. The Framework Agreement proposes to expand the definition to include certain payments that do not require production. This would allow the United States to include its counter-cyclical payments (CCPs) in the Blue Box, rather than in the AMS

[4] A joint technical-assistance program involving the FAO, International Trade Centre (ITC), UNCTAD, World Bank, WTO and UNDP was established after the first WTO Ministerial in Singapore in 1996.

[5] It is often overlooked that the central message of neo-classical trade theory is that consumers are the ultimate beneficiaries from trade liberalization as a result of the price-reducing effects of increased productive efficiency and competition. Changes in the merchandise trade balance do not necessarily correlate with the changes in national welfare that are associated with trade liberalization. For an empirical example of this see Ingco (1997)

[6] OECD. Producer and Consumer Support Estimates Database, 2004. Total Support Estimate (TSE) of $349.8 billion for OECD member countries in 2003. Database accessible through http://www.oecd.org

[7] Data from USDA, Economic Research Service, http://www.ers.usda.gov/Briefing/WTO/tariffs.htm.

[8] Van Meijl and Tongeren (2001) summarize the results of major studies on the impact of further agricultural trade liberalization in the current round of negotiations. The estimates of global welfare gains and gains to developing countries in Diao et al. (2001) are the smallest for the studies examined.

[9] The EC had contended that the Lomé accords were free-trade agreements and were covered under Article XXIV of the GATT. The Lomé accords did not receive approval by the GATT as free-trade agreements.

[10] It must be determined that countries have established or are making continual progress towards establishing the following: market-based economies; the rule of law and political pluralism; elimination of barriers to US trade and investment; protection of intellectual property; efforts to combat corruption; policies to reduce poverty, increase the availability of health care and educational opportunities; protection of human rights and worker rights; and elimination of certain child-labour practices.

[11] The WTO has two safeguard instruments. These are Article XIX, as elaborated in the Uruguay Round Agreement on Safeguards, and the special safeguard provisions (SSG) contained in the Agreement on Agriculture. Both are designed to address sudden increases in imports that cause or threaten to cause serious injury to domestic producers. The number of countries that can use the SSG and the commodities to which it can be applied is limited. In contrast, all countries are able to use the safeguards provisions.

[12] Norway and Switzerland apply the same duty-free access to LDCs as provided by the EBA. Hungary, the Czech Republic, New Zealand and the Slovak Republic provide duty-free and quota-free access to all imports from eligible LDCs. The preference schemes of new (2004) members of the European Union are now subsumed under existing EU schemes.

[13] Although the enabling decision for preferential tariff treatment only refers to its provision by developed countries it provides for 'the Contracting parties to consider on an *ad hoc* basis under the GATT provisions for joint action any proposals for differential and more favorable treatment" that do not fall under the provisions set out in the decision (GATT 1979, footnote 2)

REFERENCES

Achterbosch, T., Van Tongeren, F. and De Bruin, S., 2003. *Trade preferences for developing countries.* Lei, Den Haag. Projectcode 62760 [http://library.wur.nl/wasp/bestanden/LUWPUBRD_00320895_A502_001.pdf]

Alexandraki, K., 2005. Preference erosion: cause for alarm. *Finance and Development* (March), 26-29. [http://imf.org/external/pubs/ft/fandd/2005/03/pdf/alexandr.pdf]

Anderson, K. and Martin, W., 2005. *Agricultural trade reform and the Doha development agenda.* World Bank, Washington. World Bank Policy Research Working Paper no. 3607.

Anderson, K., Martin, W. and Van der Mensbrugghe, D., 2005. *Would multilateral trade reform benefit Sub-Saharan Africans?* World Bank, Washington. World Bank Policy Research Working Paper no. 3616.

Brenton, P., 2003. *Integrating the least developed countries into the world trading system: the current impact of EU preferences under everything but arms.* World Bank, Washington. World Bank Policy Research Working Paper no. 3018.

Bureau, J.C., Gozlan, E. and Jean, S., 2005. La libéralisation des échanges agricoles: une chance pour les pays en développement? *Revue Française d'Économie,* 20 (1), 109-145.

Diao, X., Somwaru, A. and Roe, T., 2001. A global analysis of agricultural reform in WTO member countries. *In:* Burfisher, M.E. ed. *Agricultural policy reform in the WTO: the road ahead.* Economic Research Service, U.S. Department of Agriculture, Washington, 25-40. Agricultural Economic Report no. 802. [http://www.ers.usda.gov/publications/aer802/aer802c.pdf]

FAO, 2002. *FAO papers on selected issues relating to the WTO negotiations on agriculture.* FAO, Rome. [http://www.fao.org/documents/show_cdr.asp?url_file=/DOCREP/005/Y3733E/y3733e08.htm]

GAO, 2001. *International trade: comparison of U.S. and European Union preference programs.* U.S. General Accounting Office, Washington. [http://www.gao.gov/new.items/d01647.pdf]

GATT, 1979. *Differential and more favourable treatment, reciprocity and fuller participation of developing countries: decision of 28 November 1979.* GATT, Geneva. [http://www.wto.org/gatt_docs/English/SULPDF/90970166.pdf]

Hoekman, B., Ng, F. and Olarreaga, M., 2002. Eliminating excessive tariffs on exports of least developed countries. *The World Bank Economic Review,* 16 (1), 1-21.

IMF, 2004. *World economic outlook: the global demographic transition.* IMF, Washington. [http://www.imf.org/external/pubs/ft/weo/2004/02/]

Ingco, M., 1997. *Has agricultural trade liberalization improved welfare in the least developed countries? Yes.* World Bank, Washington. World Bank Policy Research Working Paper no. 1748. [http://www.worldbank.org/research/trade/pdf/wp1748.pdf]

Mattoo, A., Roy, D. and Subramanian, A., 2002. *The African growth and opportunity act and its rules of origin: generosity undermined?* IMF, Washington. IMF Working Paper no. WP/02/158.

Subramanian, A., 2004. *Financing of losses from preference erosion.* WTO, Geneva. WT/TF/COH/14.

Tangermann, S., 2002. *The future of preferential trade arrangements for developing countries and the current rof WTO negotations on agriculture.* FAO, Rome. [http://www.fao.org/DOCREP/004/Y2732E/y2732e00.htm]

UNCTAD, 2003a. *Back to basics: market access issues in the Doha agenda.* UNCTAD, New York. [http://www.unctad.org/en/docs/ditctabmisc9_en.pdf]

UNCTAD, 2003b. *Trade preferences for LDCs: an early assessment of benefits and possible improvements.* UNCTAD, New York. [http://www.unctad.org/en/docs//itcdtsb20038_en.pdf]

UNCTAD, 2004. *The least developed countries report 2004.* UNCTAD, New York. [http://www.unctad.org/en/docs/ldc2004_en.pdf]

Van Meijl, J.C.M. and Van Tongeren, F.W., 2001. *Multilateral trade liberalisation and developing countries: a North-South perspective on agriculture and processing sectors.* Agricultural Economics Research Institute (LEI), The Hague. [http://www2.lei.wur.nl/publicaties/PDF\2001\6_xxx\6_01_07.pdf]

Wainio, J., Shapouri, A., Trueblood, M., et al., 2005. *Agricultural trade preferences and the developing countries.* Economic Research Service, U.S. Department of Agriculture, Washington. Economic Research Report no. 6. [http://ers.usda.gov/publications/err6/err6.pdf]

WTO, 2001. *Ministerial declaration, 4th WTO Ministerial conference, Doha, Qatar, November 9-14, 2001.* WTO, Geneva. [http://www.wto.org/english/thewto_e/minist_e/min01_e/mindecl_e.htm]

WTO, 2004. *Doha work programme: decision adopted by the General Council on 1 August 2004.* WTO, Geneva. [http://trade.ec.europa.eu/doclib/docs/2004/august/tradoc_118356.pdf]

CHAPTER 8

IMPROVING MARKET ACCESS IN AGRICULTURE FOR THE AFRICAN LEAST DEVELOPED COUNTRIES

Deepening, widening, broadening and strengthening trade preferences

WUSHENG YU[1]

Associate Professor, Food and Resource Economics Institute, The Royal Veterinary and Agricultural University, Frederiksberg C, Denmark

INTRODUCTION

African least developed countries (ALDCs) have enjoyed preferential treatment in exporting their agricultural products to developed countries. Reducing agricultural trade barriers at the multilateral level may erode the benefits of these preferences. Therefore, any analysis on the potential impact of the Doha outcome on the ALDCs needs to consider the possibility and the extent of erosion of agricultural trade preferences.

After having implemented unilateral trade reforms and experienced multilateral trade liberalization resulting from the Uruguay Round, many LDCs are not keen on being active players in the Doha Round. Instead, they ask for further preferential treatment from developed and advanced developing countries, and exemptions from reforming their own policies. Indeed, these concerns have been included in the negotiation agenda. For instance, in the recent July Package of the WTO agricultural trade negotiations (WTO 2004), it is stipulated that "developed Members, and developing country Members in a position to do so, should provide duty-free and quota-free market access for products originating from least developed countries".

There have been ongoing debates on the desirability and feasibility of adopting this proposal. Some worry about the inability of preferences in promoting agriculture exports and economic development in the LDCs and doubt the value of preferences as an effective measure of special and differential treatment. Poor export performance of the LDCs has often been cited in supporting this argument. These worries are compounded by the fear that the preferential approach may slow down

N. Koning and P. Pinstrup-Andersen (eds.), Agricultural Trade Liberalization and the Least Developed Countries, 129–152.

the multilateral liberalization process and compromise or delay potential gains from a freer multilateral trading system. Others point out that developing countries in general could gain more from market access reforms based on the 'Most Favored Nation' (MFN) approach and that the erosion of preferences (due to MFN liberalization) does not appear to be a big issue if substantial MFN reforms are conducted multilaterally. Still others argue that the LDCs do not necessarily gain from multilateral trade reforms, that the existing preferences are vital to their interests, and that any enhancement of such preferences would help mitigate any adverse effects from multilateral reforms. Lastly, many have noticed that various conditions, clauses and rules attached to existing preference programs may have hindered recipient countries from taking full advantage of these programs and therefore preferences should not be held responsible for the poor export performance of the LDCs. Instead of giving up on preferences altogether, some argue that efforts should be made to improve the rules associated with the preference programs to make them more effective.

Taking the July Package text as the departure point, this chapter examines empirically the value of existing preference programs in agriculture to the ALDCs and investigates the merits of enhancing these programs in the current negotiations. Specifically, Section *"The debate on trade preferences: a survey"* of the paper surveys and synthesizes several recent studies on the utilization of agricultural trade preferences for the purposes of gauging the perceived value of preferences to the recipient countries. Having concluded that preferences have been utilized to a great extent, the possibility and the extent of preference erosions in the presence of further multilateral trade liberalization are then analysed. Based on these, Section *"How can market access for the ALDCS be improved through trade preferences?"* argues that there is indeed a need for improving market access in agriculture for the ALDCs through improving and strengthening trade preferences. The July Package text is interpreted as a call for deepening, widening and broadening agricultural trade preferences targeting the LDCs, including the African ones. Further, strengthening preferences is recommended as an integrated part of the proposal. The feasibility of this proposal is evaluated against the current market access barriers facing the LDCs in the preference-granting countries. Section *"A numerical evaluation of deepening, widening and broadening trade preferences"* uses a general equilibrium model to evaluate numerically the extent of possible preference erosion and to illustrate the likely consequences of adopting the proposal on improving agricultural market access for the ALDCs in the Doha Round. The last section provides conclusions and final remarks.

THE DEBATE ON TRADE PREFERENCES: A SURVEY

Existing preference programs were often established for the purposes of promoting exports from the recipient countries. By stimulating exports from these countries, it has been hoped that economic growth in the recipient country would follow. In principle, this kind of measure compensates producers in recipient countries for their high-cost production by creating a wedge between the preferential barriers and the corresponding MFN barriers imposed on exports from countries not enjoying such preferences, which is named 'preference margin'. The magnitude of the benefits obtainable from such preferences relies on the size of the preference margin and the distribution of the associated rents.

The debate on preferences not only concerns the associated short-term commercial value but also on their long-run implications concerning export-led economic growth. Moreover, as this favourable treatment as measured by the preference margin is not meant to be constant and permanent – the reference MFN trade barriers in agriculture have been reduced following the Uruguay Round and are expected to fall further as an outcome of the Doha Round[2] – the wisdom of lobbying for this intrinsically temporary favour has also been questioned.

Have preferences stimulated trade and economic development in the LDCs?

Judging from the poor export and general economic performance of the LDCs – whose export share in total world trade has actually declined since the inception of preference programs – it seems that preferences have not realized their declared purposes. However, it would be difficult to pinpoint the causes of the poor performance solely on trade preferences and simply to declare the demise of such programs. Any empirical analysis would have to establish a counterfactual scenario in which these economies had faced the MFN trade barriers and then compare this hypothetical scenario with reality. But to establish such a scenario is difficult, if not impossible. Indeed, much of the debate remains theoretical and speculative[3].

Inama (2004) observed under-utilization of several trade preference programs (covering both agriculture and non-agriculture products) by the so-called QUAD countries (USA, Canada, the EU and Japan). The study argues that the value and effectiveness of the preferences available to LDCs' exports are discounted by the observed low utilization rates. It concludes that in order to improve the utilization of existing preferences programs, in addition to expanding product coverage of such programs (especially those of the US and Japan), it is important to change the attached rules of origin to make it less burdensome for the LDCs to comply with such rules[4].

Unlike the Inama study, a recent OECD study (2004) focuses exclusively on the utilization of agricultural preferences granted by the EU and the US. The distinct feature of the study is that it takes into account the fact that exports from beneficiary countries may be eligible for multiple preference programs (i.e. multiple eligibility). Indeed, by taking this into consideration, contrary to other studies, preference utilization rates are actually quite high for both the EU and the US preference programs. For the EU programs, the overall utilization rate exceeded 89% in 2002, and half of the eligible imports that did not use preferences entered into the EU by mostly duty-free quotas and tariff suspensions. For the US programs, the utilization rate was 88% in 2002. Some of the eligible exports entered the US market under MFN rates due to rules of origin and compliance costs, whereas other eligible exports opted for the available low MFN rates.

While drawing the conclusion that the US and EU agricultural preference programs have actually been utilized to a great extent, the OECD study also points out that in comparison to the substantial trade flows under the EU programs, trade volumes under the US programs were quite small, especially those from the African countries. It suggests that the issue of low export volumes is not so much associated with the utilization of existing programs, but rather more related to the limited product coverage of such programs and to the difficulties in meeting safety and sanitary standards attached to the programs. It also notes that rules of origin are probably not a huge issue for agricultural products, as compared to the more processed products.

The issue of low export volumes under the US programs has also been examined by Wainio and Gehlhar (2004). Through detailed analysis of a dataset obtained from the US International Trade Commission, they found that many products important to the LDCs are not covered by any US preference program. Further, the MFN tariff rates for many products covered by the preference programs are quite low, thereby making the preference margins very small. Although the second observation leaves not much room for the LDCs to gain special advantages at present, the first does imply that widening product coverage may help stimulate exports from the LDCs in the future.

In summary, by taking note of the multiple eligibility phenomenon, it appears that agricultural trade preferences have indeed been utilized, implying that there are commercial values derivable from these programs. The main problem associated with these programs is the observed low export volumes[5], which is partially related to the limited product coverage of existing preference programs. Of course, the domestic constraints in the recipient countries are also to blame. Therefore, it appears that improving trade preferences by enlarging product coverage of such programs, and by asking more countries to grant such preferences, has the possibility of expanding exports from the recipient countries. In addition, simplifying rules, making the preference more stable and creating an enabling environment for investment can also go a long way towards promoting exports from the recipient countries.

Is enhancement of existing preference programs necessary? Preference erosion and multilateral liberalization

In addition to the apparent usefulness of preferences to the recipient countries, the case for enhancing agricultural trade preferences can be further argued for by analysing the issue of preference erosion in the presence of multilateral liberalization.

Is preference erosion a legitimate concern? How large would the negative effect associated with preference erosion be? In order to answer these questions, it is necessary first to discuss the basic mechanism of preference erosion before looking at the evidence. The ALDCs typically receive more favourable market access treatment in their traditional export markets than other developing countries. MFN reforms by preference-granting countries or multilateral MFN reforms erode these preferences through two channels. By definition, MFN trade liberalization reduces preference margins, thereby eroding the advantages enjoyed by the ALDCs over their competitors. Moreover, liberalization actions by the preference-granting countries will likely lower their high domestic prices and further hurt the high-cost exporters from the preference-receiving countries[6,7].

Several recent studies have discussed the impact of multilateral liberalization or MFN liberalization by individual preference-granting countries on preference-receiving countries.

Wainio and Gibson (2004) point out that the exact impact of MFN tariff cuts by the US on countries receiving its non-reciprocal preference programs depends on the scope of the preferential treatment granted, the size of preference margins, and the depth of the MFN tariff cuts. Their results show that for countries highly dependent on preferences, the negative effects of preference erosion outweigh the positive effects of MFN tariff liberalization, whereas for countries that are not as dependent on preferences, MFN tariff cuts by the US bring about positive effects, and the larger the MFN cuts, the higher the benefits as measured in increased exports. Overall, the beneficiary countries of the US preference programs would gain from MFN tariff liberalization. The study does not provide a breakdown of the effects for individual African LDCs or for these countries as a group. Therefore, it is unclear if they would be better or worse off from the MFN liberalization. Nevertheless, it does confirm that preference erosion would be an issue for those who are dependent on preferences.

A slightly later study by Wainio and Gehlhar (2004) provides a detailed description of US non-reciprocal preference programs, covering eligible products and countries, margins of the preferences (as compared to the MFN rates), products excluded from the preferences and the applicable MFN rates, and the export patterns of the beneficiary countries in the US market. Based on this detailed data analysis, the study examines whether beneficiaries of US non-reciprocal trade preference programs gain more from cutting MFN rates on products excluded from these programs or lose more from the erosion of the preferences that they do enjoy. They conclude that developing countries as a whole would gain market shares in the US market from substantial MFN tariff liberalization, and that it is counterproductive for these countries as a group to oppose MFN liberalization. In drawing this conclusion, they emphasize the potential gains from liberalizing those products that are not included in the preference programs. However, their results also show that there would be only very minor export expansions in the US market for the LDCs (Tables 7-9 in Wainio and Gehlhar 2004) and their share in total US imports would drop after MFN reforms, therefore confirming the likely vulnerable position of the LDCs in the upcoming multilateral trade liberalization.

Unlike the US preferences, which have incomplete coverage for agriculture and food products, the EU preferences granted to the LDCs provide broader product coverage and have recently been enhanced with the adoption of the EBA initiative, a move granting full duty- and quota-free market access to all the LDCs. Yu and Jensen (2005) assess the impact of the EBA initiative on the ALDCs and show that further multilateral trade liberalizations may erode the EBA preferences. Due to its limited improvement (in terms of product coverage) from previous preferences programs, welfare impacts of the EBA on the ALDCs are shown to be small. Moreover, these small gains are likely to disappear if the EU conducts MFN trade-policy reforms, resulting in an actually worse-off situation for the ALDCs. Extending the analysis to a multilateral trade liberalization scenario reinforces the above results that the LDCs may well lose due to preference erosion and higher world-market prices for their imports. These results are echoed in Bureau et al. (2004). They find that the implementation of the 'Harbinson proposal' would lead to a slight welfare gain (0.3%) for the poorest countries. However, the gain is not evenly spread – Sub-Saharan African countries as a whole would experience a slight loss (0.1%), due to preference erosion and higher costs for imported food.

Because of the differences in the EU and US preference programs and the narrower focus on the ALDCs by the Yu and Jensen study, the above-cited studies reach different policy implications. While the Wainio and Gehlhar study illustrates that MFN reforms would lead to more gains in the US market to developing countries as a whole and that multilateral liberalization is generally a better option for developing countries, the Yu and Jensen study concludes that the ALDCs may well lose from this process. However, it appears from the results of the Wainio and Gehlhar study that the LDCs' share in the US import market would decline following the MFN reform, a point that is consistent with Yu and Jensen. The Wainio and Gibson study provides indirect support to this point as well by concluding that countries highly dependent on trade preferences may lose from preference erosion. Therefore, there seems to be some agreement on the LDCs' vulnerability in coping with MFN market access reforms conducted by preference-granting countries.

HOW CAN MARKET ACCESS FOR THE ALDCs BE IMPROVED THROUGH TRADE PREFERENCES?

As the evidence gathered above suggests that agricultural trade preferences have been widely used and that preferences erosion is a legitimate concern from the perspective of the ALDCs (if not for developing countries as a whole), the next logical question to ask is how the preferential treatment – as part of the special and different treatment stipulated in the Doha Development Agenda and the July Package – can be improved.

Deepening, widening, broadening and strengthening agricultural trade preferences[8]

First, following the July Package text, developed countries can 'deepen' their preference programs by granting the ALDCs duty and quota free market access to all agricultural products that are covered in existing programs. Second, developed countries can 'widen' the coverage of their preference programs by extending the duty and quota free access to those agricultural products that have not been covered in existing programs. These two types of actions essentially imply EBA-style preference programs by all developed countries to the ALDCs. Third, preferential market access for ALDC exports can be 'broadened' to include advanced developing countries in the group of preference-granting countries. Lastly, preference-granting countries can 'strengthen' existing preferences programs and new preference initiatives by making them permanent and unconditional. One possibility is to develop a set of WTO rules that would be applied to all preferential programs targeting the LDCs. Among the new rules, there should be simpler rules of origin and minimum administrative costs to the exporters. There should be no places in the rules for any eligibility conditions (except the one for being a LDC) that would exclude certain LDCs from the program and safeguard clauses that may discourage the recipient countries from investment and from gaining substantial market shares.

Having proposed the above measures to improve market access for the ALDCs, the next question to explore is whether there is scope for implementing the proposal, given the current state of existing preferences programs. The answer here is an emphatic yes. Broadening preferences by including advanced developing countries is possible as these countries generally have not yet provided the LDCs extensive and substantial trade preferences. Strengthening existing trade preferences is also feasible as there are many problems associated with individual programs that limit their effectiveness in promoting exports from the recipient countries. The possibility of deepening and widening preferences granted by developed countries, however, deserves some elaboration.

Existing preference programs of the EU, the US and Japan, and scopes for further improvement

In the case of the EU, there seems to be limited room for improving its preference programs because of the recently adopted EBA initiative. Upon fully implementing the EBA (including the phasing out of the transitory measures for sugar, banana and rice), the EU will be in a good position to argue for EBA-style preference from all developed countries and advanced developing countries[9].

The cases of the US and Japan are quite different from that of the EU. There, deepening and widening preferences for the ALDCs will require meaningful actions. In the case of the US, this requires expanding the coverage of the existing programs to currently excluded products. For the Japanese programs, this implies both expanding the product coverage and deepening the preference margins for the covered products.

The ALDCs receive preferences from the US through the GSP program for the LDCs, which is typically more favourable (duty-free access to covered exports) as compared to that for the non-LDC countries. Many ALDCs have also become eligible for the African Growth and Opportunity Act (AGOA). Data from the USITC data web[10] show that out of around 1800 US tariff lines, about 400 MFN tariff lines are duty-free. Among the remaining tariff lines, about 1100 lines are duty-free for the LDCs through the US preference programs. However, these preferences only lower the simple average tariffs faced by the LDCs marginally (from the overall simple average of 9.7% to the simple average of 5.6% for the GSP-LDC countries). This is because the dutiable tariff lines not covered in the preference programs generally have higher tariffs than those lines covered in the preference programs. Therefore, there is scope for extending those preferences to the dutiable lines that are not covered in the current US preference programs.

Like the US, Japan grants preferences to the LDCs through the GSP program[11]. Before 2003, this program granted preferences to around 300 tariff lines (out of around 2000 lines) for the LDCs, reducing the average duty for the LDCs from 15.6% to 14.2%. For those lines that are not covered by the GSP, there are around 400 duty-free lines and more than 1300 dutiable lines. Those uncovered dutiable lines generally have higher tariff rates. Unlike the US GSP program, the Japanese GSP programs did not grant duty-free access for the covered products and the average tariff rate for covered products were 9.8% for the LDCs, only slightly lower than the average preferential rate for non-LDC GSP countries. Since 2003, Japan expanded the GSP product coverage for the LDCs by adding around 200 products or about 10% of total tariff lines. So, it seems that Japan would have to make extensive concessions to the LDCs in order for them to enjoy universal duty and quota free access to its market.

In addition, there is also ample room for strengthening these preferences. The chapter by Blandford provides a long list of difficulties associated with the implementation of existing preferences, ranging from eligibility, product coverage, rules of origin, certainty of commitments, to the number of schemes.

Take the recent EBA initiative as an example. The safeguard measures specified in the GSP of the EU are largely retained in the EBA, with some amendments. Most notable among the amendments is the addition of the situation of 'massive imports into the EU market' as a trigger for withdrawing the preferences. With regard to the three sensitive products (sugar, bananas and rice), the EU is allowed to suspend the preferences entirely if imports cause serious disruptions to the EU's mechanisms that regulate these products[12]. In addition, the rules of origin specified in the GSP also apply to the EBA initiative. Likewise, the US and Japanese GSP program also contains various preconditions and clauses. According to the USITC data web (www.usitc.gov), the preferences offered through the AGOA are meant for all 48 Sub-Saharan African countries but until recently only 37 countries from this region have gained eligibility. Likewise, only 41 LDCs are deemed eligible for its GSP-LDC preferences. The Japanese GSP program also contains safeguard clauses and there is a graduation clause to exclude one country's exports from the program when they reach certain market share and certain minimum value.

These measures and preconditions are clearly detrimental to creating a stable trading environment for the ALDCs, and it may discourage producers in the ALDCs from committing needed investment for the purposes of reducing their high production cost. As pointed out by Panagariya (2002), it is exactly due to these measures that preferences have been rendered ineffective. It is foreseeable that strengthening the legal status of these preferences by making them universal, permanent and binding by WTO rules could well boost their performance.

A NUMERICAL EVALUATION OF DEEPENING, WIDENING AND BROADENING TRADE PREFERENCES

Methodology and data

In this section, hypothetical scenarios of deepening, widening and broadening agricultural trade preferences for the ALDCs are conducted using a numerical simulation framework. Due to the multiregional and multisectoral nature of the issues to be analysed, such scenarios are simulated within a global computable general equilibrium model named GTAP (Hertel 1997, chapter 2). The GTAP model is a standard global trade model that features intersectoral linkages through nested Constant Elasticity of Substitution production functions, and international trade linkages through the Armington specifications (Armington 1969). The demand side of the model is featured by a Constant Difference Elasticity demand system. Standard neoclassical assumptions such as constant returns to scale, perfect competition, profit and utility maximization are applied in the model. With this modelling structure, explicit welfare analysis and decomposition pertaining to a change in trade policy can be conducted.

The GTAP model is accompanied by a global data set commonly known as the GTAP database (Dimaranan and McDougall 2002). The main components of the database are detailed input–output tables for all the regions/countries included in the dataset, consistent bilateral trade flows among all the regions/countries, a protection data set that covers *ad valorem* tariff equivalents[13], export subsidies as well as domestic support measures, and macroeconomic aggregates. Together, these components give a snapshot of the world economy at the base year of the database and all the usual equilibrium conditions are satisfied in the database.

The latest version of the database, version 6, contains data for 86 regions and 57 commodities for the year 2001, including a fairly detailed breakdown of agricultural and food products. This study applies an aggregated version of this most recent GTAP database with 21 aggregated regions and 24 aggregated products. The six individual African LDCs (Malawi, Mozambique, Tanzania, Zambia, Madagascar and Uganda) are included in the aggregated database as one group (with the abbreviation of SSA-1), whereas other African LDCs are largely included in the aggregated Rest of Sub-Sahara African (SSA-2) regions[14]. Among the non-LDC regions are influential agricultural trading countries/regions such as Australia and New Zealand, China, Japan, India, Canada, the US, Argentina, Brazil and EU-25. Agriculture and food products in the original GTAP database are incorporated in the aggregated version as separated ones, including: paddy rice, wheat, cereal grains, vegetables and fruits, oil seeds, plant fibres, other crops, other animal products, bovine meats, other meats, vegetable oil, dairy, processed rice, sugar, other processed food products, and beverages and tobacco. In addition to these, non-agricultural products are aggregated as natural resources, textile and clothing, manufacturing and services.

Scenarios

The deepening, widening and broadening scenarios can be formulated as reduction/removal of relevant tariffs facing exporters from the ALDCs. In this study, the GTAP version-6 database is viewed as the initial equilibrium point of the world economy. By applying the shocks pertaining to the policy scenarios to the model, new equilibria after these shocks will be computed and updated datasets corresponding to and describing the new equilibria will then be generated. The differences between the original dataset (the base case) and the updated datasets can then be summarized and viewed as the effects attributable to the policy changes.

Three hypothetical scenarios are considered. Scenario 1 is a multilateral market access liberalization scenario in which all the non-LDC regions contained in the aggregated dataset are assumed to halve their MFN tariff rates for all agricultural and food products. To be consistent with the July Package proposal, the ALDCs are not assumed to conduct any reductions in their own market access barriers. Such a scenario sets a benchmark against which the subsequent broadening and deepening scenarios can be compared.

Scenario 2 is the deepening and widening scenario. In addition to the MFN market access reforms as simulated in Scenario 1, advanced economies (Australia and New Zealand, Japan, rest of East Asia – mainly Korea and Taiwan –, Canada, the US and the EU-25) are assumed to deepen and widen their preferential treatment for the ALDCs to the extent that all tariffs imposed on exports from the ALDCs are eliminated. This is essentially to assume an EBA offer from all advanced countries. The original initiator of the EBA (the EU-25) is assumed to implement the EBA in its entirety, implying that the transitory measures on sugar, rice and bananas are to be removed immediately. For other advanced countries, this scenario implies widening product coverage of their respective preference programs and deepening preference margins for covered products. Since the shocks contained in Scenario 1 are also included in Scenario 2, the differences between results obtained from Scenarios 1 and 2 can then be attributed to the deepening of trade preferences.

Scenario 3 is the broadening scenario. The design of this scenario again allows for comparison with the previous scenarios. Here, both the multilateral market access reform shocks and the deepening shocks are included, in addition to the new shocks involving extending EBA-style preferences to the ALDCs by several large developing economies, including China, India, Mexico, Argentina, Brazil and ASEAN (the Association of South-eastern Asian Nations).

Preferential tariff rates facing the African LDCs

Before proceeding to the simulation results, it is necessary to discuss an adjustment made to the GTAP protection data, which are aggregated from the more detailed MacMaps dataset at HS-6 levels, using bilateral trade weights. This practice, nevertheless, causes serious problems in correctly measuring market access barriers facing the ALDCs. As the ALDCs have either very few or no exports under many tariff lines (see Appendix Tables 1 and 2), the actual protections are greatly underestimated by the trade-weighted tariffs, which in many instances are simply zero. This is certainly not correct, considering the fact that preference programs in countries such as the US and Japan exclude many dutiable products, hence exposing the ALDCs to generally high MFN rates in those products. Thus, the trade-weighted aggregation scheme fails to capture the actual protection faced by the ALDCs[15]. It also leaves little room for implementing the broadening and deepening scenarios, which involves cutting the MFN rates to the preferential levels. Moreover, if these tariffs were used in simulating the above scenarios, the degree and extent of preference erosion due to multilateral liberalization would also be underestimated because cuts to preference margins relative to the initial preference margins implied by any MFN reform would be smaller with the trade-weighted tariffs (as the starting point) than it should be.

One way to remedy the downward bias associated with the trade-weight method is to apply a simple average scheme – which does not use trade flows as weights – to recalculate aggregate tariffs on exports from the ALDCs, based on the detailed source data from MacMaps. Owing to the fact that there are usually only a few tariff lines appearing for any individual ALDC country in MacMaps, taking the simple averages on a bilateral basis would lead to an incomplete representation of the barriers facing individual ALDCs. Therefore, in calculating the simple averages, tariff lines at the HS6 levels imposed on all ALDC are pooled together, with the assumption that for any given export destination, all ALDCs face the same import barriers[16]. This treatment can be justified by observing that the ALDCs are typically grouped together under existing preference programs and generally face the same preferential and MFN tariffs in a given market. As such, a certain tariff line recorded for one ALDC but not for another can very well be the applicable rate for the latter, should the latter start to export under that line. By way of the above procedure, a better representation of trade barriers facing the ALDCs, including the existing preferential tariff rates, is obtained. The original GTAP database is modified to reflect these changes and the modified database serves as the starting point for the simulations.

Results

Simulation results from the three policy scenarios are summarized in Tables 1 and 2. Here the focus is on the changes in total exports from the two ALDC regions and the resulting changes in economic welfare measured in equivalent variations. To facilitate discussing the individual effects of multilateral market access reforms, the deepening and widening of preferences and the broadening of preferences, results for Scenario 1 are calculated as changes/percentage changes from base case data, whereas results for Scenario 2 (Scenario 3) are computed as changes/percentage changes from the updated dataset obtained from Scenario 1 (Scenario 2). In other words, the results reported for Scenario 2 are due to the deepening and widening of preferences only, whereas the results reported for Scenario 3 are due to the broadening of preferences only.[17]

Scenario 1

As can be seen from Table 1, total agricultural and food exports from both SSA-1 and SSA-2 would drop by over six% under Scenario 1. Underlying this aggregate change are near-universal declines in exports of all agricultural and food products. The largest percentage changes are in vegetables and fruits, bovine meats, other meats and sugar. However, the most significant changes in terms of trade volumes are in other crops, other food, and vegetables and fruits, as these are the products in which the two ALDCs have substantial base case exports. For instance, the decreases in exports of other crops of 5.9% for SSA-1 and 4.4% for SSA-2 are equivalent to losses of export volumes of around US$ 60 million for the former and US$ 150 million for the latter. Among the few exceptions to this declining pattern are the slight increases in exports of rice and plant fibres. However, only the increases in plant fibres seem to be meaningful as the base case exports of rice are very small.

Based on these results, it seems that the two African regions would lose part of their exports in the wake of the assumed multilateral market access reforms, provided that no further preferences are granted. The decline in their exports is coupled with increased world trade in virtually all agricultural and food products. In fact, total world export volume in dollar terms would be boosted by almost 6% due to the market access reform, implying that the ALDCs' shares of agriculture exports would shrink.

Scenario 2

Deepening and widening trade preferences by developed countries would reverse the negative export effects on the two ALDCs created by the multilateral market access reform. Results from Scenario 2 (also in Table 1) show that as compared to Scenario 1, total exports of agricultural and food products from SSA-1 would increase by over 17% whereas those from SSA-2 would increase by around 30%. In dollar terms, following the deepening and widening action, total agricultural and food exports from SSA-1 would be over US$ 2.4 billion, representing an increase of over US$ 360 million from Scenario 1. For SSA-2, the increase is almost US$ 2.5 billion. These increases more than make up for the losses sustained from the multilateral market access reform.

The increase in total agricultural exports would not be evenly distributed across products. Those products that are important to the ALDCs and that are excluded from the current preference programs are the ones that would experience the greatest increase. In percentage terms, the increases are the highest for meat products, dairy products and sugar for both regions. In addition, exports of vegetables and fruits and oil seeds would also increase significantly for SSA-2. Most notable among the changes are the increased exports of sugar, reaching over US$ 400 million for SSA-1 and around US$ 2.5 billion for SSA-2, due to the high market access barriers for non-LDC exporters (hence, large preference margins) and that several advanced countries have maintained substantial trade barriers for sugar exports from the ALDCs.

Table 1. *Changes in exports of selected agriculture and food products from SSA-1 and SSA-2*

	Scenario 1				Scenario 2				Scenario 3			
	Export volume				Export volume				Export volume			
	(million US$)		% change		(million US$)		% change		(million US$)		% change	
	SSA-1	SSA-2	SSA-1	SSA-2	SSA-1	SSA-2	SSA-1	SSA-2	SSA-1	SSA-2	SSA-1	SSA-2
grains	35.8	48.4	-2.6	-9.5	36.3	48.6	1.3	0.4	36.3	49.1	0.1	0.9
vegetables & fruits	130.0	810.1	-10.9	-8.3	130.2	1109.0	0.2	36.9	212.3	1192.5	63.0	7.5
oil seeds	25.2	236.1	0.7	-21.5	25.5	403.7	1.2	71.0	25.2	398.8	-1.3	-1.2
plant fibres	112.9	896.5	0.3	0.5	108.3	841.8	-4.1	-6.1	113.3	910.7	4.6	8.2
other crops	1040.6	3322.0	-5.9	-4.4	1078.7	3047.9	3.7	-8.3	1137.6	3104.8	5.5	1.9
bovine meats	1.6	25.8	-26.0	-14.4	1.9	24.4	17.6	-5.5	2.1	25.0	12.6	2.7
other meats	7.5	36.4	-11.4	-9.4	9.7	53.8	28.9	47.6	14.6	61.6	50.5	14.5
vegetable oils	6.8	142.7	-8.4	-7.8	6.5	130.0	-4.5	-8.9	6.8	129.4	5.5	-0.5
dairy	1.1	29.3	-8.0	-5.8	3.3	47.9	191.6	63.5	3.3	49.1	1.1	2.6
rice	5.7	30.9	2.3	1.8	5.5	28.8	-3.7	-6.6	5.3	28.5	-3.8	-1.1
sugar	93.6	169.9	-21.3	-49.8	411.6	2507.8	339.9	1376.1	402.6	2499.4	-2.2	-0.3
other food	508.3	2058.1	-5.6	-5.6	513.7	2034.3	1.1	-1.2	511.4	2079.8	-0.4	2.2
Total agri-food	2052.1	8226.2	-6.5	-6.9	2415.3	10697.5	17.7	30.0	2555.7	10960.3	5.8	2.5
Total	6569.2	52985.5	-0.7	-0.5	6674.8	53948.0	1.6	1.8	6718.2	54071.2	0.7	0.2

Sources: simulation results

***Table 2**. Welfare results for selected countries/regions (million US$)*

	Scenario 1			Scenario 2			Scenario 3		
	Efficiency	Terms of trade	Total	Efficiency	Terms of trade	Total	Efficiency	Terms of trade	Total
Australia & New Zealand	4.2	566.2	545.3	1.2	5.8	6.1	-0.2	-6.5	-6.8
China	830.8	-164.6	575.2	3.9	-1.6	-4.4	3.0	-6.1	-2.6
Japan	3263.5	-536.2	2766.7	-4.3	-37.3	-49.3	2.0	-0.7	0.5
Rest E. Asia	1141.5	-32.5	1067.6	-81.5	-27.0	-109.8	2.1	-0.9	1.1
ASEAN	554.7	289.6	760.9	-2.6	9.8	7.1	-7.5	-24.1	-31.2
India	830.8	-216.0	610.9	-6.9	-8.1	-16.6	22.5	-26.9	-4.6
Canada	674.5	-112.9	551.3	1.4	9.7	11.0	0.3	-2.4	-1.8
USA	87.5	957.0	1292.6	14.5	-68.3	-99.4	2.1	-10.7	-18.4
Mexico	321.2	-208.1	108.9	-0.1	6.3	5.8	4.2	-3.5	1.0
Argentina	46.3	259.4	270.1	-0.1	-2.4	-2.2	0.0	-1.8	-1.6
Brazil	162.2	888.0	1098.6	-3.0	-5.4	-10.3	1.2	-5.0	-4.3
EU25	5586.4	-1255.9	4276.2	-72.6	-494.1	-582.3	15.5	-33.0	-19.9
SSA-1	-1.4	-42.5	-50.0	3.9	91.4	111.8	-1.1	48.2	53.5
SSA-2	-38.9	-126.8	-184.2	168.2	527.9	772.4	17.0	66.4	91.2
World	16400.2	-12.2	16387.8	3.6	-4.6	-1.6	62.5	-0.2	62.3

Sources: simulation results

In contrast to the large export expansions of many products, exports of several products from the two ALDCs would decrease. Notable examples are exports of plant fibres from both regions and other crops from SSA-2. This is due to the intersectoral resource movement triggered by the expansion of preferential coverage and the deepening of existing preference programs. In fact, the existing preferences may have distorted production and trade patterns in the beneficiary countries. Making such preferences universal and homogenized across sectors may help the beneficiary countries reconfigure their production and trade patterns according to true comparative advantages so as to avoid narrow or wrong specialization. For instance, the expansion of exports of 'other food products' would lead to declining exports of 'other crops' (mainly tropical products) in SSA-2. This in turn may also help mitigate the long-term trend of declining prices of such products.

Scenario 3

Those developing countries chosen for conducting the broadening scenario (Scenario 3) generally do not offer extended preferential treatment targeting the ALDCs, their imports from the two African regions are very small, and in some cases no such imports exist according to the GTAP database. So the resulting changes in exports from the ALDCs in Scenario 3 not only depend on the MFN market access barriers of the chosen developing countries, they are also related to the initial export volumes from the ALDCs. The latter is important since the modelling framework adopted for this paper uses the so-called Armington trade structure, which is known to experience difficulties in generating trade when there is little or no trade to begin with.

The overall increase in agricultural exports due to the broadening of trade preferences would be around US$ 130 million for SSA-1 and US$ 260 million for SSA-2. The main sources of such an increase are from vegetables and fruits, plant fibres, other crops, and meat products. In contrast, exports of sugar, rice and oil seeds from both regions actually decrease slightly. It should be noted that the overall increases in exports reported for Scenario 3 are much smaller than those obtained from the deepening and widening scenario (Scenario 2). Although this result may have something to do with the Armington trade structure employed in the model and the fact that there is little agricultural trade between the ALDCs and those developing countries (that are assumed to grant preferences), the market size of the developed countries and their role as the ALDCs' traditional markets may be more responsible for the relatively larger export effects from the deepening and widening scenario[18]. This result seems to discount the optimism on the South-South trade, at least in the short and medium run.

Welfare effects[19]

While the multilateral market access reforms would benefit most non-LDC countries, the welfare effects turn out to be negative for the two African regions (losses of about US$ 50 million and 184 million for SSA-1 and SSA-2, respectively), a result that is consistent with Yu and Jensen (2005).

To understand these welfare results from Scenario 1, the focus should be on the negative export price effect, which dominates the total terms-of-trade effect for both regions. This negative export price effect is due to two reasons. On the one hand, multilateral market access reforms would lead to lower prices in the export markets, and hence lower prices for those ALDC exports covered in preference programs. At the same time, lowering MFN market access barriers would lead to higher prices for exports from countries not receiving preferential treatment. Hence, non-LDCs would be able to export and crowd out exports originated from the ALDCs. On the other hand, preferential access granted to the ALDCs would actually 'trap' their exports and prevent them from shifting to other markets, thereby further dampening the prices of ALDCs' exports. In addition, the ALDCs may be also hurt by higher world market prices for their imports.

The negative welfare effects on the two African regions would be more than offset by the deepening of existing preference programs of the developed countries. Results from Scenario 2 show that such a move by the developed countries would not only result in improved terms-of-trade for the African LDCs, it would also lead to efficiency gains for them. For SSA-1, the total welfare improvement from Scenario 1 would be over US$ 110 million, whereas for SSA-2 this would be almost US$ 800 million. Most of these gains are due to improved terms-of-trade, with the positive export price effects being the dominant factor.

While deepening preferences by the developed countries seems to generate substantial benefits to the African LDCs, according to the simulation results, broadening preferences would not generate similar exports expansion and welfare gains to the African LDCs. The additional welfare gain to SSA-1 from the broadening scenario would be a little over US$ 50 million and that to SSA-2 would be around US$ 90 million.

Effects on preference-granting and other countries

Deepening trade preferences by developed preference-granting countries would lead to small terms-of-trade losses to these countries. For instance, the EU-25 would suffer a welfare loss of US$ 582 million (see Table 2). However, this loss is much smaller than the gains obtained from the multilateral market access reforms (i.e. Scenario 1). On balance, the developed countries would still gain significantly from multilateral market access reforms, even if taking into consideration their losses from deepening their preference programs to the ALDCs. For non-LDC developing countries, the negative impact of more favourable preferential treatment for the ALDCs would also be very small, implying that the expansion of exports from the ALDCs would generally not be a big concern for them. For example, the economic welfare of China and India would be reduced by only about 4 and 17 million US dollars, respectively. As such, by broadening trade preferences to the ALDCs, the advanced developing countries would suffer very minor welfare losses as well.

Overall, the cost of broadening and deepening preferences for African LDCs appear to be very minor to other countries. Although not presented here, the trade diversion effects are also very small, a result that is consistent with the ALDCs' very small exports in total world trade. Therefore, the concern on trade diversion does not appear to be a big issue.

CONCLUSIONS

The July Package of WTO agricultural trade negotiations calls for duty and quota free access for exports from the LDCs. This chapter discusses the merits of this proposal.

The usefulness of preferences has been revealed by the high utilization rate of agricultural trade preferences. The case for improving trade preferences is further supported by several recent studies examining the possibility and extent of preference erosions. Based on these, this chapter proposes deepening, widening, broadening and strengthening trade preferences for the ALDCs. The feasibility of the proposal is reflected in the incomplete coverage of existing preferences programs in key countries and in the complicated conditions and rules attached.

A set of CGE simulations illustrates the potential impact of this proposal. The first policy scenario confirms the ALDCs' vulnerability in multilateral trade liberalization, in terms of reduced export volumes and export shares and deteriorated terms-of-trade. Deepening and widening trade preferences by developed countries (Scenario 2) would more than offset these negative effects on the ALDCs. At the same time, harmonizing the preferences programs to duty-free and quota-free access for all products by developed countries would help reveal true comparative advantages of the ALDCs. Adding selected advanced developing countries to the preference-granting group (i.e., broadening preferences, as in Scenario 3) would further expand exports from the ALDCs and improve their economic welfare. It should be noted that the added benefits from broadening preferences would be smaller than what could be achieved from the deepening and widening scenario. This result emphasizes the importance of free access to developed countries' markets and differs from the belief that enhanced South-South trade may benefit the LDCs more.

Of course, these benefits would not be fully realized without strengthening the legal foundation of the preference programs. The ALDCs need to conduct domestic policies reforms aiming at creating an enabling environment for their export-oriented industry to take advantage of this opportunity.

While the current paper provides some support to the July Package text on offering duty and quota free access to exports from the LDCs, the political feasibility of this proposal is entirely another matter. Nevertheless, the numerical results of the chapter suggest that the proposal would impose little cost on the rest of the world due to limited trade diversion. Moreover, one need not worry about the conflict between the preferences and the multilateral liberalization process. The very reason that some African countries are not willingly participating in the Doha Round is partly due to their fear of losing out on the preference front. An offer of deepening, widening, broadening and strengthening preferences should create the right incentive for them to agree to a new deal. And this narrow yet vital interest of the ALDCs will by no means jeopardize the whole dynamics among major trading nations, and implementing this idea will not alter the world trade patterns.

Having argued for improving preferences for the ALDCs and having illustrated numerically the benefits to these countries and the small costs to preference-granting countries, a cautionary word should be added. Just as one should not dismiss the value of the preference programs for their poor historical performance, one also needs to realize the limits and diminishing nature of this favourable treatment. Preferences cannot and should not be viewed as a source of competitiveness. Rather, they only provide an important yet temporary opportunity for the ALDCs to expand and develop their economy. Over-estimating the value of preferences is just as misleading as not granting this opportunity or not taking advantage of it.

NOTES

1 I thank David Blandford, Niek Koning and Per Pinstrup-Andersen for suggestions on pursuing this topic and for valuable comments on earlier drafts of the paper. Useful discussions with Mark Gehlhar, Alan Matthews, Chantal Nielsen and John Wainio are acknowledged. I am also indebted to Betina Dimaranan for providing detailed tariff data. The views expressed in this paper are mine alone and should not be attributed to the institute with which I am affiliated.

2 Unless preferential market access barriers continue to fall at the same rate as MFN barriers, multilateral reforms will inevitably erode the preferences margin. As preferences for some products have already reached duty- and quota-free status, any MFN reforms will definitely lead to preference erosion for these products. This indeed points to the nature of such programs – they are meant to be temporary and exporters from the LDCs are expected to become competitive when the preference margin ceases to exist.

3 The chapter by Blandford provides a more detailed discussion on the general role of preferences.

4 Brenton (2003) provided an initial evaluation of the impact of the EBA for the year 2001 and found out that utilization of this initiative by non-ACP LDCs was low. The study suggests that the rules of origin may be to blame. However, the fact that the study only used data gathered for the first year of implementing the EBA and the limited effective product coverage of the EBA may also explain the low utilization rate found in the study.

5 This problem can be illustrated using data on bilateral exports from the African LDCs. Appendix Tables 1 and 2 provide such an overview. There, export volumes smaller than US$1 million are shaded. A casual look reveals two features of export patterns of these countries: very low export volumes and very narrow export concentrations.

6 An example of this point is the reform of the Common Agricultural Policy (CAP) of the EU. A study by Frandsen et al. (2003) shows that reforming the EU sugar policy may hurt the recipient countries of tariff-free sugar quotas and benefit more efficient exporters that do not receive preferences.

7 In addition to the erosion caused by market access reforms, possible negative terms-of-trade effects caused by removing agricultural subsidies in the OECD countries are also a concern for net food-importing LDCs. Lowering these subsidies will likely reduce the incentives for farmers to overproduce in the OECD countries and will lead to higher world-market prices for food and agricultural products. Moreover, many LDCs have already had difficulties in keeping their balance of payments in check. These price shocks will likely exacerbate the situation. Lastly, these negative effects may well be compounded and reinforced by the many domestic supply-side constraints (which prevent them from taking full advantage of any export opportunities arising from trade reforms) and the chronic external debt burdens (which make it difficult for them to finance more expensive imports) of these countries. Some of these points have been addressed in Yu and Jensen (2005) in their analysis of the EBA initiative of the EU.

8 Blandford (2004) proposed these measures and argued that they could help improve the effective participation of the LDCs in the multilateral trading system.

9 One nuance is that the EU may need to balance the interests of different types of recipients of its preference programs. For example, the transitory measure adopted for sugar exports from the LDCs

may be more a response to the demands from non-LDC ACP countries than to those from domestic producers in the EU.

[10] These are drawn from the summary compiled by Breton and Ikezuki (2005) and Wainio and Gehlhar (2004).

[11] Numbers in this paragraph are calculated from the UNCTAD TRAINS database and are drawn from Breton and Ikezuki (2005).

[12] Serious disruptions refer to, among other things, reduction in market shares of European producers, reduction in their production, increases in their stocks, closure of their production capacity, bankruptcies, low profitability, low rate of capacity utilization, employment, trade and prices (EC 2001a; 2001b).

[13] The newest GTAP version-6 database incorporates market access barrier data contained in the MacMaps data set (Bouët et al. 2004), which encompasses not only *ad valorem* tariff rates, but also *ad valorem* equivalence of specific tariffs and Tariff Rate Quotas.

[14] The aggregated SSA-2 region contains 43 individual countries, 33 of which are LDCs and the rest are non-LDCs. The GTAP version-6 database does not provide a further breakdown of this region. Therefore we are forced to treat this as an aggregated LDC region. Any preference granted by developed and advanced developing countries in practice and in the hypothetical scenarios of the study is assumed to be available to the non-LDC countries in SSA-2 region as well. Consequently, numerical results obtained for this aggregated region are for both the LDC members and non-LDC members of this group. Nevertheless, as the majority of countries in this group are LDCs and most of the non-LDC members also receive preferences, it is expected that this is a meaningful grouping.

[15] A rather extreme example to illustrate this point is to consider the Japanese rice tariff: while the trade-weighted tariffs facing other exporters range from 300% (for the EU-25) to 1000% (for China), they are simply zero for the ALDCs!

[16] Of course, the ALDCs face different barriers in different export destinations.

[17] The results presented in this section are computed without including the Japanese tariff on rice in the deepening and widening scenario. This warrants some explanation here. The Japanese rice tariff is set at a prohibitive level. In the multilateral market access scenario, the assumed halving of this tariff would result in a new tariff that is still prohibitive. Meanwhile, a complete deepening scenario would remove this tariff for the two African regions. As a result, exports – and hence outputs – of rice in the two regions would increase dramatically, leading to massive resource reallocation into rice production. However, considering the size of the Japanese rice market, it is not really credible for Japan to maintain a prohibitive tariff on everybody else but the ALDCs. As such, in the scenarios reported here, this possibility is excluded.

[18] A simple sensitivity analysis with respect to the Armington elasticities has been carried out by re-running the three experiments with a new set of elasticities that are twice as large as the original ones used in the GTAP model. Results from these simulations show that the increases in agricultural exports from the African LDCs will be higher under both the deepening and the broadening scenarios, as compared to those reported in Table 1. Nevertheless, higher Armington elasticities boost exports under the deepening scenario much more than under the broadening scenario, suggesting that the qualitative conclusion reported in the main text is quite stable with respect to the degree of substitution in the Armington structure.

[19] Results discussed here are comparative static aggregate welfare effects, obtained by calculating the equivalent variation. These results can not be directly used to evaluate the effect of trade policy changes on poverty. But it is well established in the literature that farm export expansion – which has been observed in the results of this paper – has important multiplier effects for economic development in the poor countries.

REFERENCES

Armington, P., 1969. A theory of demand for products distinguished by place of origin. *IMF Staff Papers*, 16 (1), 159-178.

Blandford, D., 2004. *Failure to achieve progress in the WTO: how large are the benefits foregone by the LDCs? paper presented to the workshop on Agricultural Trade Liberalization and the Least Developed Countries, Wageningen, The Netherlands, December 2-3 2004.*

Bouët, A., Decreux, Y., Fontagné, L., et al., 2004. *Computing an exhaustive and consistent, ad-valorem equivalent measure of applied protection: a detailed description of MAcMap-HS6 methodology*. CEPII, Paris.

Brenton, P., 2003. *Integrating the least developed countries into the world trading system: the current impact of EU preferences under everything but arms*. World Bank, Washington. World Bank Policy Research Working Paper no. 3018.

Brenton, P. and Ikezuki, T., 2005. The impact of agricultural trade preferences, with particular attention to the least developed countries. *In:* Aksoy, M.A. and Beghin, J.C. eds. *Global agricultural trade and developing countries*. World Bank, Washington.

Bureau, J.C., Gozlan, E. and Jean, S., 2004. *La libéralisation du commerce agricole et les pays en développement: conférence internationale "Les politiques agricoles sont-elles condamnées par la mondialisation?" jeudi 7 octobre 2004, Paris*. CEPII, Paris. [http://jcbureau.club.fr/ifri_17sept04.pdf]

Dimaranan, B.V. and McDougall, R.A., 2002. *Global trade, assistance, and production: the GTAP 5 data base*. Center for Global Trade Analysis, Purdue University, West Lafayette.

EC, 2001a. Council Regulation (EC) No 416/2001 of 28 February 2001 amending Regulation (EC) No 2820/98 applying a multiannual scheme of generalised tariff preferences for the period1 July 1999 to 31 December 2001 so as to extend duty-free access without any quantitative restrictions to products originating in the least developed countries. *Official Journal of the European Communities,* L60, 43-50. [http://trade.ec.europa.eu/doclib/docs/2004/october/tradoc_111459.pdf]

EC, 2001b. Council Regulation (EC) No 2501/2001 of 10 December 2001 applying a scheme of generalised tariff preferences for the period from 1 January 2002 to 31 December 2004. *Official Journal of the European Communities,* L346, 1-59. [http://www.delkhm.cec.eu.int/en/special_features/EBARegulation.pdf#search=%22council%20regulation%202501%2F2001%22]

Frandsen, E.F., Jensen, H.G., Yu, W., et al., 2003. The EU sugar policy: an analysis of price versus quota reductions. *European Review of Agricultural Economics,* 30 (1), 1-26.

Hertel, T.W., 1997. *Global trade analysis: modeling and applications*. Cambridge University Press, Cambridge.

Inama, S., 2004. *Trade preferences for LDCs: a quantitative analysis of their utilization and suggestions to improve it: paper presented to the 7th Annual Conference on Global Economic Analysis, Washington, DC, June 17-19*. [https://www.gtap.agecon.purdue.edu/resources/download/1722.pdf]

OECD, 2004. *Assessment of the utilization of selected preferences in the EU and US agricultural and food markets: joint working paper on agriculture and trade, Directorate for Food, Agriculture and Fisheries and Trade Directorate*. OECD, Paris.

Panagariya, A., 2002. EU preferential trade arrangements and developing countries. *The World Economy,* 25 (10), 1415-1432.

Wainio, J. and Gehlhar, M., 2004. *MFN tariff cuts and US agricultural imports under nonreciprocal trade preference programs: paper presented at the 7th GEA Conference, Washington DC, June 17-19*.

Wainio, J. and Gibson, P., 2004. The significance of US nonreciprocal trade preferences for developing countries. *In:* Anania, G., Bohman, M.E., Carter, C.A., et al. eds. *Agricultural policy reform and the WTO: where are we heading?* Edward Elgar, Cheltenham.

WTO, 2004. *Doha work programme: decision adopted by the General Council on 1 August 2004*. WTO, Geneva. [http://trade.ec.europa.eu/doclib/docs/2004/august/tradoc_118356.pdf]

Yu, W. and Jensen, T.V., 2005. Tariff preferences, WTO negotiations and the LDCs: the case of the'Everything but arms' initiative. *The World Economy,* 28 (3), 375-405.

Appendix Table 1 *Base case export volumes of SSA-1 (million US$)*

	Australia and New Zealand	China	Japan	Asean	India	Canada	USA	Mexico	Argen-tina	Brazil	EU25	Mid-East and N.Africa	S.-Afr. Custom Union	SSA-1	SSA-2	World
Paddy rice	0	0	0	0	0	0	0.2	0	0	0	0.5	0	0	0.1	1.3	2.5
Wheat	0	0	0	0	0	0	0	0	0	0	0	0	0	0.3	0.3	0.6
Grains	0.3	0.4	0.9	0.6	0.1	0.5	3.3	0.2	0.2	0.1	8.1	0.6	0.2	10.5	9.3	36.7
Fruits and vegetables	0.5	0.6	3	3.9	53.6	1.3	6.7	0.3	0.2	0.3	56.8	5.6	2.7	1.2	4.8	145.9
Oil seed	0	0.1	11.2	0.4	0.3	0.3	0.8	0	0	0	4.8	1.2	2.9	0.6	1.5	25
Plant fibres	0	3.4	0.7	18.3	13.8	0.1	0	0.1	0	0	39.3	1.4	12.3	0.6	7.2	112.6
Other crops	11.1	2.5	70.1	105.3	3.7	2.9	115	10.6	1.8	2.1	505	50	22.3	34.6	79.1	1106.1
Other animal products	0.1	1.9	1.3	2	1	0.2	3.5	0.2	0	0	10.9	2.8	1.1	1	1.9	50
Wool	0	0	0	0	0	0	0.1	0	0	0	0.2	0	0	0	0	0.5
Bovine meats	0	0	0.1	0.1	0	0.1	0.4	0	0	0	0.9	0.1	0	0	0	2.2
Other meats	0.1	0.1	0.3	0.2	0	0.1	1	0.1	0	0	2.6	0.2	0.1	1.2	1.6	8.5
Vege oils	0	0	0.1	0.9	0	0	0.3	0	0	0	0.5	0	1.7	0.9	2.3	7.4
Dairy	0	0	0	0.1	0	0	0.1	0	0	0	0.1	0.1	0.1	0.2	0.5	1.2
Rice	0	0	0	0.1	0	0	0.3	0	0	0	0.8	0.1	0.1	0.4	3.3	5.6
Sugar	0	0	0	0	0.1	0	16.7	0	0	0.5	37.3	7	1	10.8	44.3	118.8

App. table 1 (cont.)

App. table 1 (cont.)

	Australia and New Zealand	China	Japan	Asean	India	Canada	USA	Mexi-co	Argen-tina	Brazil	EU25	Mid-East and N.Africa	S.-Afr. Custom Union	SSA-1	SSA-2	World
Other food	0.7	1.3	39.3	12.4	0.2	1	11.4	0.4	0.3	0.2	393	6.2	10.7	3.3	29.5	538.7
Beverage / Tobacco	0.3	0.3	1	0.4	0	0.5	3.4	0.2	0.1	0.1	8.4	0.6	1.2	0.3	3.3	22.8
Total ag. food	13.2	10.7	128.5	145.1	72.8	7.3	164.8	12.2	2.7	3.3	1073	76.3	56.4	66	190.2	2193.7
Total	28.8	104.3	290.6	290	124.4	36.5	643.4	26.6	10.2	14	3206.8	268.8	576.5	130.9	285.4	6614.2

Sources: GTAP database version 6.

Note: For presentation purposes, numbers smaller than 1 million US dollars are shaded in the table.

Appendix Table 2. *Base case export volumes of SSA-2 (million US$)*

	Australia & New Zealand	China	Japan	Asean	India	Canada	USA	Mexico	Argentina	Brazil	EU25	Mid-East and N. Africa	S Afr, Custom Union	SSA-1	SSA-2	World
Paddy rice	0	0	0	0	0	0	0.1	0	0	0	0.2	0	0	0.1	7.4	7.9
Wheat	0	0	0	0	0	0	0	0	0	0	3.7	1.3	0	0.2	2.5	8.1
Grains	0.3	0.7	1.3	0.6	0.1	0.6	4.4	0.2	0.2	0.2	12.6	4.7	0.1	0.2	12.1	53.5
Fruits and vegetables	0.3	0.6	5.3	2.9	60.7	0.5	8.2	0.4	0.1	0.4	724.9	17.6	0.8	0.3	40.1	883.6
Oil seed	0	0.1	31.3	1.1	0.1	0.9	6.9	0	0	0	65.2	102.7	1.3	0.3	6	300.5
Plant fibres	0.3	3.2	3.4	194.7	107.5	2.8	0.4	7.4	0	32.3	261.5	73.4	2.5	0.3	33.4	891.8
Other crops	6.8	10.9	83.8	34.1	11.6	45.4	346.8	6.7	0.9	18.1	2195.1	225.6	9	10.6	151.5	3474
Other Animal products	0.1	3.4	1.7	7.7	12.3	0.6	7.7	0.9	0	0.1	74.7	7.5	0.4	1.1	5.8	154.1
Wool	0.2	0.3	0.9	0.4	1.7	0.3	2.4	0.2	0.1	0.1	5.6	0.4	0.1	0	0.1	14.2
Bovine meats	0.1	0.2	0.3	0.1	0	0.1	1.1	0.1	0	0	3.4	22.7	0.1	0.5	0.5	30.1
Other meats	0.2	1	0.6	0.5	0.1	0.5	2.4	0.1	0.1	0.1	16.5	1.5	0.3	3.2	11	40.2
Vege oils	0.1	0.2	0.8	0.2	0	0.2	8.9	0.1	0	0	86.9	0.3	0.1	4.1	45.7	154.8
Dairy	0.1	0.1	0.2	0.5	0	0.1	0.7	0	0	0	9.1	1.6	0.1	1.4	16	31.1
Rice	0.2	0.4	0.1	0.5	0.1	0.4	2.9	0.2	0.1	0.1	5.8	0.5	0.1	0.1	17.5	30.3

App. table 2 (cont.)

App. table 2 (cont.)

	Australia & New Zealand	China	Japan	Asean	India	Canada	USA	Mexico	Argentina	Brazil	EU25	Mid-East and N. Africa	S Afr, Custom Union	SSA-1	SSA-2	World
Sugar	0.3	0.2	0.2	0.4	0.1	0.2	15.2	0.1	0.1	0.1	295	1.4	0	0.5	17.6	338.6
Other food	9.3	28.9	104.6	59.4	1	8.4	69.7	4.6	1.3	1.9	1465.1	22.2	14.1	20.1	234.5	2180
Beverage/Tobacco	0.8	1.1	6.1	1.7	0.3	2.1	15.6	0.8	0.4	0.4	42.1	2.3	0.4	4.9	85.7	173.6
Total ag. food	19.4	51.7	242	305.3	195.7	63.5	497.3	22	3.4	53.9	5277.3	493.2	30	47.9	728.9	8835.8
Total	144.4	2281.5	1271.5	1226.3	680.4	405.4	14625.9	218.2	109.5	1453.9	21176.1	1300.2	1537.2	381.7	2118.9	53253

Sources: GTAP database version 6.

Note: For presentation purposes, numbers smaller than 1 million US dollars are shaded in the table.

CHAPTER 9

AGRICULTURAL TRADE LIBERALIZATION UNDER DOHA

The risks facing African countries

OUSMANE BADIANE

Africa Coordinator and Senior Research Advisor to NEPAD;
International Food Policy Research Institute (Washington, DC), USA

INTRODUCTION

Global agricultural policies affect many economies in a similar way, but may have more severe economy-wide consequences in Africa. Most African countries find themselves at the lower spectrum of the economic development process, which implies a greater dependence for overall economic growth on domestic demand, agricultural incomes and agricultural trade. In addition, the structure of domestic production and export sectors, the level of capacities to absorb economic shocks, and the historically outward-oriented nature of the economies, when taken together, constitute distinguishing characteristics of African countries at present. These features do not only make African economies more vulnerable to distortions and changes in global trading policies in the agricultural sector, but they also determine the implications of agricultural trade liberalization among African countries.

The first section of the paper discusses relevant features of African economies and examines the resulting vulnerability with respect to global agricultural trading policies and their induced changes in world agricultural markets. The second section presents some of the key outcomes of the Uruguay Round Agreement on Agriculture and their implications for African agriculture. It also highlights lessons learned and future global trade policy challenges and options facing African countries. The final section of the paper looks at the ongoing agricultural trade negotiations, identifies potential risks for African countries and discusses options for global trade liberalization that would best benefit African economies.

N. Koning and P. Pinstrup-Andersen (eds.), Agricultural Trade Liberalization and the Least Developed Countries, 153–173.

The paper will argue that trade preferences have not been beneficial to African economies, have not compensated them for the negative impact of global protectionism, and are unlikely to do either in the future. Moreover, it will show that the insistence on the part of African countries on Special and Differential Treatment entails much more risk than benefit for their economies. The paper will also disagree with the widely accepted conclusion that African countries would suffer from liberalization of global agricultural policies because they tend to be net food importers. This conclusion does not sufficiently take into consideration the dynamic long-term effects of global policy changes on production and trading patterns among African countries and the potential efficiency effects that would emanate from them.

RELEVANT FEATURES OF AFRICAN ECONOMIES

Long before the debate about economic openness occupied the centre of the post-structural-adjustment growth and development agenda, economists started to stress the critical link between overall trade, economic development and growth performance in the agriculture sector in the early stages of the development process[1]. The growth literature of the 1960s and 1970s in particular emphasized the importance of domestic demand for the growth process. Its findings suggest that "a minimum threshold of development is needed before export growth and economic growth are associated" (Heller and Porter 1978, p. 192), and that the weak relationship between the two in the early period of development is due to the "relatively low level of manufactured exports in several countries" (Balassa 1978, p. 183). Furthermore, the analysis of the relative contributions of domestic and foreign demand to economic growth by Urata (1989) shows a much stronger contribution of domestic absorption at lower levels of economic development. A key conclusion from the above is that, at lower levels of development, the stimulus for structural transformation and growth must come from internal demand, which in turn is fuelled by growth in the agricultural sector.

The crucial role of agricultural growth as a stimulus to the process of overall growth has also been documented in micro-level studies. For instance, in their study of small enterprises in several African countries, Liedholm et al. (1994) found that differences in local agricultural growth were the most important determinants in explaining the differences in enterprise start-up rates and expansion, as well as in employment creation in the studied zones. Similarly, a study by Delgado et al. (1994) on growth linkages within the local economy in a sample of African countries estimated growth multipliers to be much larger than previously thought and fully comparable to estimates in the Asian literature. The estimates obtained in their study show the effect of adding one dollar to farm tradable incomes in the study zones to be an increase in total incomes by 2 to 3 dollars. In other words, a sustained expansion of revenues from agricultural tradables would result in an increase in overall incomes in the local economy that is at least twice as high as the initial increase of incomes in the agricultural sector itself.

Globalization has introduced a significant change in the growth dynamics implied above. Falling transport costs, development in international finance, higher levels of trade exchange with the rest of the world and greater degree of openness of domestic economies, have together gradually reduced the dominant role of domestic demand in stimulating structural transformation and growth. These factors explain why Africa's situation today is very different from that of Asia in the 1950s and 1960s. The greater role of internal demand combined with lower levels of external competition in domestic markets in Asia during that period meant that supply-raising agricultural technology advances could go a long way towards meeting part of the growth challenge. This, in principle, explains the extent of the success and impact of the green revolution. African countries find themselves today in a different situation. Globalization and its associated factors listed above mean that advances on the supply side are more intricately linked to factors on the demand side. African countries do not only have to produce more, they also have to 'sell' better in far more competitive domestic as well as external markets in order to raise supplies. The agricultural sector still operates as a crucial stimulant of structural transformation and growth. However, the growth of the sector itself depends on factors outside the domestic economy and the supply-side sphere. Chief among the factors affecting agricultural and overall growth among African countries are global protectionism and its associated policies, including explicit and implicit export subsidies and dumping.

In sum, the vulnerability of African countries with respect to global policies and the trade liberalization agenda arises from characteristics inherent to their economies, such as: strong dependence on agriculture for income, employment and foreign exchange earnings; low shock absorption capacity at national as well as household levels; heavy dependence on food imports; and relatively high degree of sector openness. These conditions render African economies particularly vulnerable to trends and instability levels of world agricultural prices, long-term changes with respect to access barriers to export markets, and global policies affecting the competitiveness of imports in domestic markets across Africa. They also determine the cost to African countries of current global policies as well as eventual gains from trade liberalization.

AFRICAN COUNTRIES AND GLOBAL TRADE LIBERALIZATION: LESSONS FROM THE URUGUAY ROUND AGREEMENT ON AGRICULTURE

The Agreement on Agriculture has not adequately addressed Africa's needs

The objective here is to review the outcomes from the past trade negotiation round and their impact on African countries as a first step towards examining the opportunities and risks facing them under the Doha agenda. There is now a consensus that the Agreement on Agriculture (AoA) has not led to any significant reform of global agricultural policies. In fact, a closer look at the evidence would reveal a marked deterioration in several areas. Figures 1 and 2 below summarize trends in overall support to agriculture, price protection and export subsidies among OECD countries. It is clear from the different graphs that support to agriculture has indeed grown since the agreement in 1994, as has the level of protection as measured by the ratio of farm-gate prices attributable to border protection.

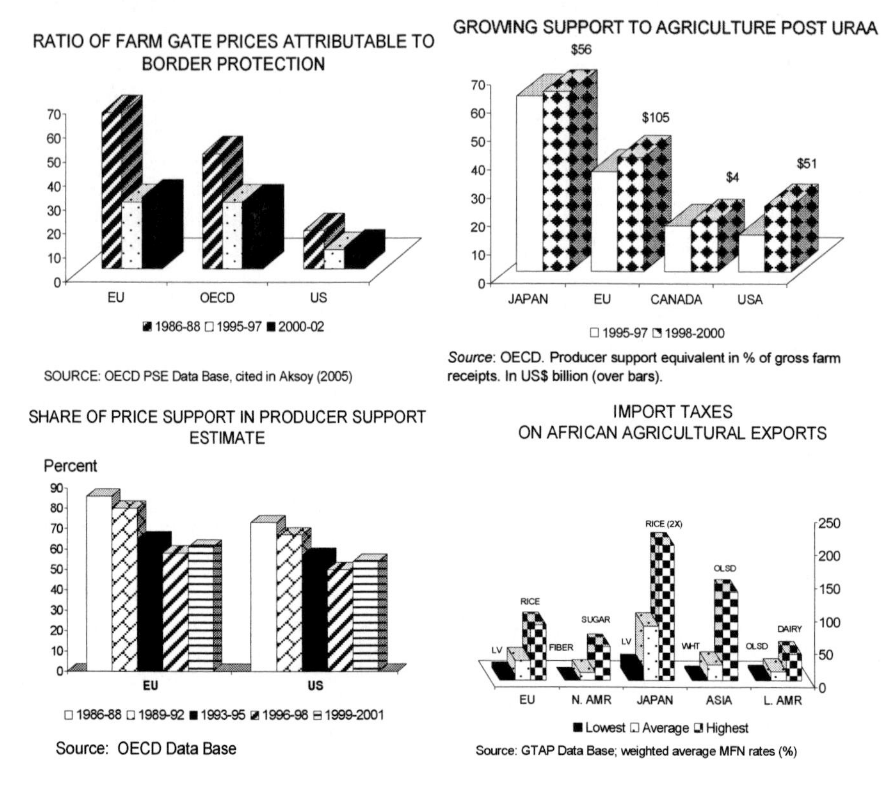

***Figure 1**. Trends in OECD policies*

Moreover, the use of price as a main instrument of support is still widespread. Both in the US and EU, the extent of price support and its share in total producer support have increased again after the mid-1990s. To be sure, they remained lower than in the base years 1986-88. As is well known and documented, that base period was a real outlier and by selecting it, opportunity was provided to weaken the agreement. High levels of protection and support are also prevalent among developing countries, as documented in Anderson et al. (2005).

In the case of African countries, the graphs show that African exports indeed face import taxes in all regions and that these taxes can be quite high for some exports. At the same time, exports to African countries continue to be heavily subsidized.

While it is true that African countries enjoy considerable preferences, in particular when exporting to the EU, the increase in global protection and continued use of price as an instrument of support mean that the associated distortionary implications for world markets in terms of price levels, structure and stability, have remained if not amplified. These effects are transmitted directly into African economies and shape the environment for production and consumption in these countries. The ramifications emanating from this affect the performance of their agricultural sectors. The situation is being further complicated by the emergence of increasingly complex norms and standards and other types of non-tariff barriers. Jaffee and Henson (2005) and Wilson (2002) illustrate the considerable challenge facing African countries with respect to complying with the quality requirements for agricultural exports.

Preferences may not be working

It is often stated in the literature that African countries would lose from further liberalization of global policies given that they are already enjoying preferences that would be greatly eroded by further tariff reduction. This could have been true if the preferences were working. However, a closer look at the evidence reveals that, for whatever reason, they may not be effective. An extensive analysis of the effects of preferences on agricultural exports by African countries has been carried out by Brenton and Ikezuki (2005). The graph in the top right-hand-side corner of Figure 2 is based on their results. It shows the benefit of current preferences by the EU, Japan and the USA to African countries. For that purpose, the value of preferences is expressed in terms of the share in the total exports of individual preference-receiving countries. The countries are then regrouped in three different groupings depending on whether the value of preferences is less than 1% of exports or contained within the 1-5 or 5-20% ranges. In the graph, the bars represent the number of preference-receiving African countries that fall within each of the three groups. For the great majority of African countries, the value of preferences is no more than a small percentage of their exports. The value of preferences granted by the US and Japan, for instance, is less than 1% of exports for about 80% of African countries. Preference benefits are highest for exports to the EU, where more than half of African countries have values ranging between 5 and 20%.

African countries are frequently blamed for failing to exploit these preferences. There may be a multitude of reasons why the value of preferences has been so low, many of them in fact linked to the global trading policy environment, whose impact preferences seek to mitigate. These would include: (i) the long-standing pressure on international agricultural prices due to the rapid expansion of subsidized production in OECD countries over the last four decades; (ii) the increasing instability of world-market prices due to the protection and isolation of domestic markets in an increasing number of countries since the 1960s; and (iii) the degree of distortion that has been introduced into world agricultural markets following decades of intervention in the agricultural sector by dominant trading partners. The agricultural sector in African countries is widely exposed to these developments, which have negatively affected its performance. Furthermore, Africa's agriculture has been constantly besieged by heavily subsidized exports from a host of sources, not only OECD countries, as shown in the top left-hand-side graph of Figure 2. The recent debates around the Doha round have highlighted the case of the cotton sector in

West Africa, which is just the latest and perhaps more prominent sector to have fallen victim to global protectionism. A closer examination would reveal similar ramifications in other regions in Africa and sectors, including oilseeds, dairy, cereals, beef and, recently, poultry.

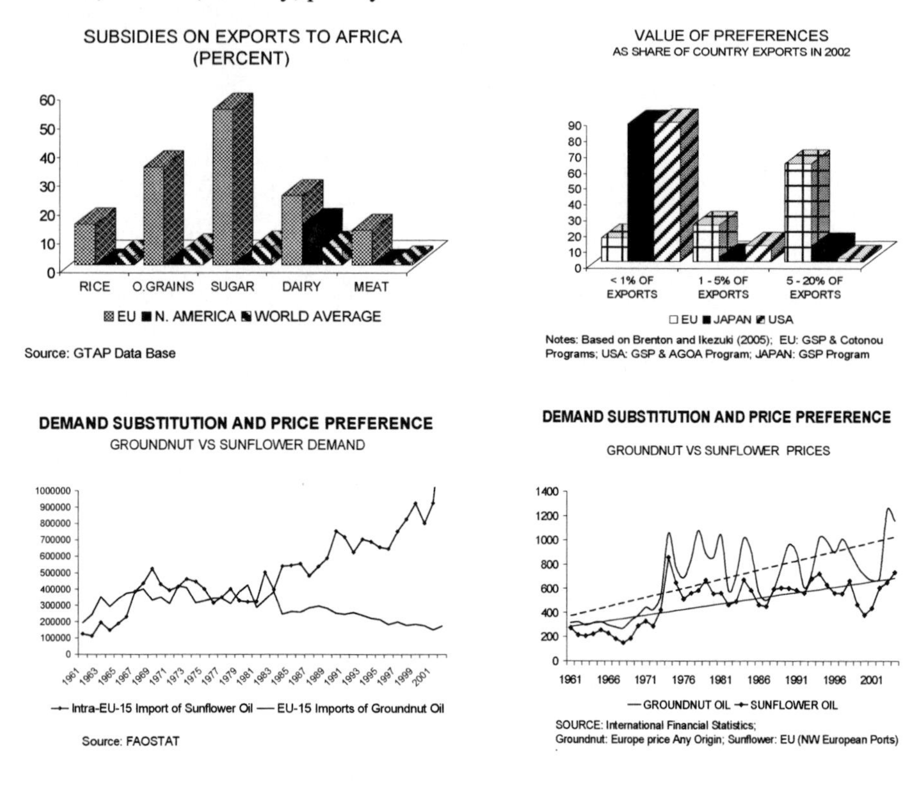

Figure 2. *Subsidies and preferences*

Demonstrative effects of global protectionism undermine preferences

No one can deny the considerable harm that countries' own sectoral policies have done to agriculture in the past. Despite nearly a decade of reforms, detrimental policies still prevail in many African countries. Although they are not necessarily motivated by policy intervention in the North, African governments currently justify distortionary sector policies by pointing to policy regimes in OECD countries, as they have done since the start of the structural adjustment related policy reforms of the 1980s. More importantly, agricultural and agribusiness interest groups more frequently ask their governments simply to copy the North[2]. Recent developments in the cotton sector in West Africa can again be cited here as example. Until the recent collapse of world cotton prices, momentum was gathering to address the institutional and policy weaknesses in the sector. Significant progress was made in the policy debate across the region and some consensus was emerging about the need to reform the sector and create more transparency, greater efficiency and increased participation of farmer organizations in decision making and management. When the world cotton market crisis hit, all the attention was suddenly turned towards subsidies and other support policies in other major exporting and/or producing countries and away from the considerable problems that were threatening long-term viability of the sector from within. It was not the first time that cotton prices collapsed, nor were the policies in competing countries being discovered for the first time. A major factor in the immediate and forceful response among West-African governments to the fall in cotton prices has been its timing. The debate about reforms was at its highest level of intensity and the pressure to move and reform was mounting. It was clear to many actors involved that difficult decisions lied ahead. The crisis was therefore seen and seized as a welcome opportunity to step back from these decisions. The risk and importance of internal institutional and policy deficiencies with respect the sector's long-term viability were quickly ignored. All efforts were now diverted towards fighting production and export subsidy policies in competing countries. The reform process was simply put on hold or eventually rolled back.

A less visible and talked about example happened just recently. At a Presidential Forum organized by an African government and attended by about ten foreign heads of state, the main recommendation was that African countries should erect tariff walls and seek to double prices paid to their farmers. Ironically, the same speakers also recommended that African governments ask developed countries to provide the necessary financial aid also to double prices of export commodities. These positions were defended by several international keynote speakers. The reason was that such policies have succeeded among developed countries. Although the above arguments are untenable and the recommendations certain to be never implemented, they were received with amazing support by the audience. The broad support among the audience highlights one of the many problems associated with the protectionist policy regimes of developed countries: their demonstrative effects for African as well as other developing countries. These regimes have provided some kind of legitimacy for interventionist policies and the continued reluctance to reform them just reinforces the position of those who see no evil in them but rather point to the phenomenal increase in output and dominant position in international export markets after four decades of massive support by developed countries. They do not bother going through the complex analysis of the significant economic losses caused to protecting countries and the high cost imposed on the global agricultural systems.

One can hardly ignore nowadays the fact that global protectionism has emerged as a credibility problem for national as well as international proponents of further reforms in Africa's agricultural sector. It has gradually eroded the support for further reforms that are necessary to restore growth in African agriculture, as it is increasingly perceived as the villain while the exorbitant costs of past domestic policy mistakes are fading in people's memory. The emerging point of view for an increasing number of stakeholders is that, if there is anything wrong with past policies, it is the fact of having reformed or abandoned them. The probable failure of the Doha round in terms of effective liberalization of global trade policies would most likely roll back some of the important sector policy improvement among African countries and hence undermine their capacity to exploit future and possibly expanded preferences.

Demand erosion, demand substitution, and price preference

Preferences basically allow recipient countries to export to preference-granting countries at higher prices. They do not offer protection against export demand erosion due to subsidized expansion in the preference-granting countries of output in the sectors for which preferences are being granted. Nor do they protect against demand substitution, which arises when protection and domestic subsidies boost output of substitute products, which then displace exports in competing sectors. The case of vegetable oil exports to the EU market can be used to illustrate the implications of demand substitution. The bottom left-hand-side graph in Figure 2 shows the evolution of extra-EU imports of groundnut oil compared to intra-EU imports of sunflower oil, a close substitute. The latter rose rapidly during the 1960s, and, by the end of that decade, surpassed the volume of groundnut oil imports, which from that period onward fell steadily. Over the following 30 years, groundnut oil imports into the EU fell by more than half, from a peak of about 400,000 metric tons in the early 1970s to less than 200,000 metric tons in 2002. Intra-EU sunflower oil imports, in contrast, more than doubled to about 1 million metric tons in the same period.

The graph to the right shows the evolution of relative prices during the same period. While the rapidly expanding demand for high-value vegetable oil in the EU has been captured by EU sunflower oil producers, induced changes in world market price ratios between the two products have gradually shifted competitiveness and demand outside of the EU in favour of sunflower oil. Moreover, instability in the groundnut oil market increased substantially over the same period, compared to sunflower oil prices, in addition to a generalized pressure on world vegetable oil prices. That pressure resulted not only from rising vegetable oil production in the EU but also from the expansion of soybean production in countries such as the US and Brazil in order to meet the expanding demand for substitute feed in the EU, following the substantial increase in protected cereal prices in that market. The world market was flooded with soybean oil, a by-product of supplying the EU with soybean meal. The consequence was not only lower export prices for African groundnut oil exporters but also increasing competition in domestic and cross-border markets among African countries. The combination of demand erosion/substitution, unfavourable shift in relative prices, generalized price decline and increasing competition in local markets does not only hurt current producers, processors and exporters, but it also significantly undermines the incentives for long-term investment in the sector. Consequently, the capacity of preference-receiving countries to fill their quotas is weakened, as is that of new entrant countries to invest in expanding production and exports. The trends in the graphs in Figure 2 would suggest that benefits from preferential access to the EU market would be quite limited, certainly much less than African exporters, such as Senegal, would have realized in the absence of the policies that induced the substitution in import demand as well as changes in world prices and vegetable oil supply.

The arguments of demand erosion and substitution are often dismissed quite hastily on the grounds that African countries have often failed to fill their quotas. The counter-argument is weakened by the fact that, like the preferences themselves, the capacity of countries to fill their quotas cannot be treated separately from the global trading environment and its consequences on production and consumption conditions in recipient countries. The failure to fill preference quotas is closely linked to the performance level of export sectors and its underlying factors. As outlined in the preceding sections, the overall performance of domestic sectors in Africa has been significantly affected, directly and indirectly, by global protectionism and trading practices. In other words, preferences are being undermined by the same distortionary effects of global protectionism that they seek to alleviate. The issues are thus broader than price preference and exemption from border protection. Price-related preference erosion should therefore not be treated separately from demand erosion, demand substitution and the possible adjustment in domestic production and consumption patterns in African countries that would result from effective global trade liberalization.

Preferences and incentives for long-term investment in agriculture

A major weakness of preferences, from the long-term growth point of view, is their concessionary character, which makes them less predictable and reliable in the long run. Consequently, they do not create enough incentives for long-term investments. Furthermore, in the context of smallholder conditions, preferences generate rents that are likely to be captured further downstream along the export supply chain, with limited incentive for farm-level investment and modernization. Moreover, in cases where exports are subject to taxation, the preference margins may end up constituting fiscal transfers from preference-granting to recipient countries, with no assurances that the resources so collected would be invested in sectors that are affected by global protectionism.

Even if broad in coverage and more predictable as in the case of EU's Everything-But-Arms (EBA) initiative, preferences would not solve the problem of demand erosion and substitution resulting from global protectionism. More importantly, the problem is bound to become more acute, the more countries shift their strategies towards markets in preference granting countries in search for ever narrowing export markets. Only effective liberalization of global policies, which is required to remove border protection and eliminate domestic support, would solve the problem of demand erosion and demand substitution. Effective liberalization would, however, render preferences unnecessary. Moreover, broad liberalization would open the rapidly expanding export markets in emerging economies for African exports. By justifying the perpetuation of global protectionism, preferences delay the access to these markets. In summary, preferences need to be coupled with broader liberalization in order to have sustained impact and affect the growth process. Broad and effective liberalization would, however, take away the justification for preferences. From the point of view of African countries, this paradox weakens the case of preferences as a strategic objective under multilateral trade negotiations. The argument of preference erosion loses appeal, unless one assumes that global trade liberalization is impossible and that demand erosion and substitution as well as international trade distortions were to persist. That assumption, albeit currently widely shared, should however not affect the way we account for the benefits and losses of the current system for African countries.

Is the current Agreement on Agriculture benefiting or hurting African countries?

As indicated above, the case against preferences and, as will be seen further below, Special Differential Treatment (SDT), as key pillars of negotiating positions for African countries, is that both are justified only in the case of continued global protectionism, which they, in turn, serve to legitimize. By definition, preferences imply the existence or – in the current case – continuation of protection. SDT for least-developed countries, on the other hand, can be seen as the *quid pro quo* for agreement with a continuation of border protection and domestic support by developed WTO member countries. The numbers shown in Figure 2 above clearly show that preferences have not compensated African countries for the harm they suffer due to global protectionism. On the other hand, the available evidence, in the majority of cases, suggests that African countries would gain from a liberalization of global policies. While there may be disagreement about the size of the gain, there is consensus about the sign of the impact[3]. In all estimates, the cost of global protectionism, as measured by the simulated changes in GDP, value added or incomes, is far greater than the estimated values of preferences reported in Figure 2.

Net food importing countries are often seen as losers because liberalization would lead to higher import prices. These simulations emphasize the effect of increasing world market prices on food import cost in these countries. They often fail to capture the dynamic effects of changes in global policy distortions on production and consumption patterns in African countries. The depressing and destabilizing effects of international agricultural policies on world market prices have been widely analysed and documented. The same policies have also distorted the structure of world market prices quite considerably and increased competition in local and trans-border markets in Africa. Furthermore, preferences and other concessionary arrangements lack the long-term predictability and reliability to induce significant investment in agriculture in African and other preference-receiving countries. Moreover, agricultural protection among OECD countries has a strong demonstrative effect among African policy makers, who see them as proof of acceptability and effectiveness of interventionist and distortionary policies.

If the Doha Round were to lead to effective liberalization, the expected changes in the structure, levels and stability of world prices and of supply conditions in domestic markets would most likely have the double effect of stabilizing and raising the average levels of profits in Africa's agricultural sector. To the extent that higher and more stable levels of profit, greater transparency in the international trading environment, and improved national policies in African countries translate into higher levels of investment and technological innovation, liberalization would accelerate the rate of growth in African agriculture in the longer run. Indeed, studies that incorporate the dynamic effects of global protectionism suggest levels of gains from global trade liberalization that are several times higher than indicated through standard comparative static methods. For instance, simulations by Anderson et al. (2005), when treating productivity endogenously, increase the gains from trade liberalization in terms of real income among developing countries from US$ 90 billion to nearly US$ 700 billion. The increase among low-income countries would be from US$16 billion to US$ 70 billion. The comparative static version of the study also indicates that the average income gain resulting from global liberalization would be higher in Sub-Saharan Africa than in all other regions. The projected agricultural output and employment growth rates are higher, or at least comparable to, rates that are estimated for other regions[4].

THE DOHA ROUND AND ITS POTENTIAL RISKS FOR AFRICAN COUNTRIES

There are two major risks associated with potential outcomes of the ongoing Doha Round negotiations: (i) a lack of effective liberalization, and thus continuation of global protectionism, with consequences similar to that of the Uruguay Round, as described above; and (ii) Special and Differential Treatment clauses which may perpetuate old, or induce new, policy distortions in African countries that are harmful to their domestic agricultural sector.

The risk of another lost decade for African agriculture[5]

From the point of view of African countries, the real risk is not whether or not an agreement will be reached under Doha. The issue is, rather, whether such an agreement will prove to be any more effective than the current AoA at reducing global policy distortions and opening up market access for African exports. There are several reasons why Doha may not lead to effective liberalization of agricultural policies in OECD countries and thus in the potentially important markets for African exports in emerging countries. Firstly, negotiation modalities for the reduction of domestic support are based on final bound levels as opposed to actual AMS levels, making it unlikely that effective reduction would take place, given the level of cuts that would be implied. Also, given the similarity of modalities with the AoA, the chances of decoupling should be limited if the experiences with the current agreement are taken as indicators. According to Baffes and de Gorter (2004), "the experience with decoupling agricultural support has been mixed while the switch to less distortive support has been uneven across commodities and countries. Rules have changed with new decoupling programs added so expectations about future policies affect current production decisions. Time limits were not implemented and if so, were overruled".

On the other hand, proposed modalities for market access foresee tariff reduction to be made from bound rates instead of actual, applied levels. In many cases, the gap between the two, or the binding overhang, can be so substantial as to render any tariff cuts ineffective. In addition, countries will have the possibility of designating 'sensitive products' which would enjoy 'flexibilities' in terms of tariff reduction. The elimination of tariff quotas is not envisaged, nor is broad reduction of in-quota tariffs on the table. It is not clear at this stage that significant progress will be achieved in terms of eliminating tariff peaks, high tariffs and tariff escalation. In addition, such efforts could be further undermined by the introduction of 'sensitive products', which may be primary candidates for the application of quotas or subject to high tariffs, tariff peaks and tariff escalation[6]. On the export competition front, export subsidies are still defined in terms of budgetary outlays and quantity commitments, as under the AoA, and not in terms of *ad valorem* subsidy equivalents. The reluctance to negotiate on the basis of *ad valorem* subsidy equivalents would lead to the same loopholes and delays in disciplining export competition. Moreover, as in the other cases, final bound commitments of export subsidy volumes and outlays are being used as a basis for further reduction, not actual levels. Although the objective is to reduce subsidies to zero by the end of implementation period, the modalities involve considerable risk of delaying subsidy cuts for important sectors.

The extent of binding overhang with respect to domestic support and export subsidy commitments is illustrated in Figure 3 below. The left-hand-side graph shows the extent of export subsidy commitment use for selected products, both in terms of volumes and outlays, by all 25 countries that are concerned under the AoA. The share of products benefiting from export subsidies is added. The graph indicates that there should be plenty of room to expand export subsidization both in terms of individual product coverage as well as aggregate quantities and expenditures. It can therefore be expected that a weak agreement would very likely fail to restore export competition effectively.

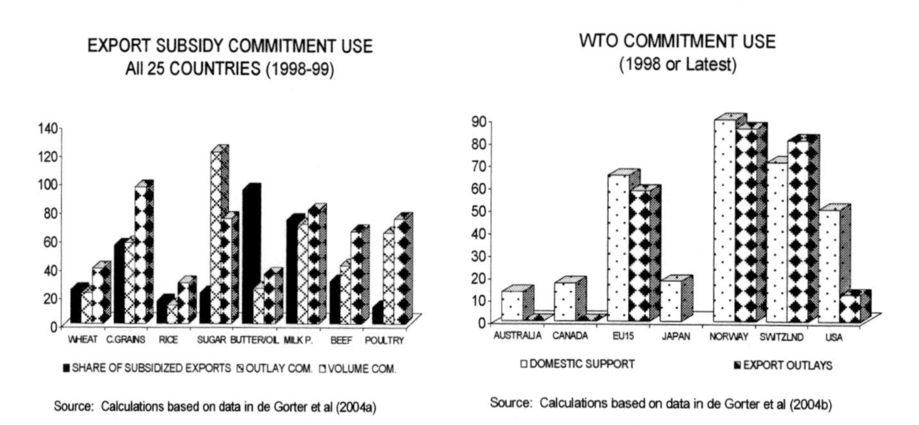

***Figure 3**. Commitment use under AoA*

The right-hand-side graph shows the situation with respect to export subsidies as well as domestic support commitment for selected exporters. Here again, it appears that the binding overhang would call for substantial cuts in domestic support and export subsidies in order to effect real changes. Failure to do so in either case would allow countries to compensate cuts in one area by expanding support in another. Still on the export side, proposed rules to discipline food aid call for the provision of non-emergency food aid in the form of untied financial grants. They would, however, allow in-kind aid to be provided within the framework of programs or projects operated by specialized United Nations food-aid agencies or non-governmental humanitarian organizations and private charitable bodies. The latter two groups would be more difficult to police and could provide considerable loopholes.

The above examples indicate that there are significant risks of another lost decade for African countries in terms of reducing global protectionism and improving access to markets. This conclusion, as well as the preceding discussion, does not ignore the difficulties of negotiating agreements nor the complexities involved in arriving at mutually acceptable outcomes. All it does is to stress the risks that these very difficulties and complexities, as reflected in the draft modalities and the work program (WTO 2003; 2004), may well mean that possible outcomes at this Round would not lead to effective liberalization of global agricultural policies over the next ten years. If that is the case, African countries have would not have much reason to expect great economic benefits from the outcomes.

SDT and its risks for African agriculture

Global negotiations are about detrimental effects of national policies upon trading partners and arrangements to reduce or eliminate such effects. They do not deal with the harmful effects of the same policies on individual countries' own domestic sectors. Agricultural policies in African countries have, however, caused more harm to domestic sectors. Furthermore, Special and Differential Treatment under global trade negotiations seeks to alleviate the burden of compliance with global policy changes among African and other low-income countries. While doing so, they not only ignore the harmful effects of national policies on local agricultural sectors, but also may perpetuate or even accentuate these effects. This risk results from the fact that SDT may be easily accepted by developed countries because they involve little cost to their economies. Ironically, such SDT may have substantial and detrimental effects on agriculture in the developing countries that are requesting them, since they often reflect more the biases of bureaucrats requesting them than the real needs of farmers in African countries.

Several proposed SDT measures are analysed below that entail the risk of encouraging policies that are detrimental to African agriculture. While warnings have been made in the past regarding the risks associated with SDT, there have not been efforts to review negotiation modalities systematically with respect to such risks. For instance, Oyejide (2002) stresses the potentially counterproductive effects of SDT. He suggests the introduction of multilateral rules governing the granting of derogation. However, he sees the risk as limited to derogation with respect to tariff reduction and is open to the granting of full derogation with respect to other obligations. As will be shown, SDT risks under Doha go well beyond tariff reduction.

Current modalities suggest that proposed disciplines outlawing new export prohibitions, restrictions or taxes on foodstuffs shall not be applicable to developing countries (WTO 2003, paragraphs 39 and 40). It is difficult to see how this derogation can be beneficial to the agricultural sector. If anything, it would legitimate and perpetuate a practice that has done and continues to do quite significant harm to the agricultural sector in African and other developing countries. Under export competition, developing countries can, under certain conditions, request an exporting country to provide more generous export-financing terms than permissible under the proposed new rules seeking to discipline export financing. This measure is open to abuse given that both exporting countries and importing countries, willing to satisfy interests of trader groups, would have incentives to use it. Similarly, SDT under export finance rules would allow developing countries to use longer maximum repayment terms and longer instalment periods for principal and interest repayments when providing export finance. While few African countries would make use of this measure, it would weaken the agreement by opening the door for continued subsidization of exports into Africa by exporting developing countries, as illustrated in Figure 2.

Proposed SDT measures under Article 6.2 would allow countries to provide input subsidies. The benefit to farmers from such derogation is obvious. However, the absence of discipline as to under which conditions and in which form such subsidies can be provided could lead to government interference in input distribution and output marketing sectors. Similarly, SDT related to provision of subsidies for concessional loans through established credit institutions for the establishment of credit cooperatives entail the risk of interference with lending policies and practices and hence viability of the banking sector. Also, SDT targeting assistance for the establishment and operation of cooperatives, risk management and compliance with sanitary and phytosanitary measures should be beneficial. It may, however, lead to intervention in the marketing system to control prices, crop movement or other sales strategies.

Under export competition, proposed rules disciplining State Trading Enterprises (STEs) would exempt developing countries from prohibition of STEs to restrict the right of any interested entity to export agricultural products or purchase such goods for export. Further, the Doha Work Program (WTO 2004) postulates that "STEs in developing country Members which enjoy special privileges to preserve domestic consumer price stability and to ensure food security will receive special consideration for maintaining monopoly status" (Annex A, paragraph 25). Another proposed rule that would be applicable to all WTO members would ensure that STEs do not export at a price that is less than the price paid to domestic producers. While such a rule would protect foreign suppliers from dumping, the exemption being granted to developing countries under the above SDT measure would, on the other hand, allow STEs in these countries to suppress export demand and thus lower prices paid to local farmers. While it is difficult to see how such derogation could benefit the agricultural sector in African countries, its risks for the sector are enormous and obvious.

In all the above cases, SDT should spell out certain principles governing their application. One may contend that global negotiations are not the place to address strictly self-inflicted harm and that it is the responsibility of SDT-using countries to use them wisely and to the benefit of their farming sector. History has taught us that this is not always the case, certainly not in many African and developing countries. On the other hand, it can be argued that SDT measures resulting from global negotiations provide some sense of global legality to distortionary practices in the sense that they can be now seen as WTO compliant. When practices are being cemented and legitimized in international agreements, it is certainly justified to expect that such agreements provide safeguards against abuse of these practices. For instance, disciplines could be introduced in connection with the above SDT, which would ensure that subsidies are applied at farm level and without interference with the pricing and distribution of inputs by private sector operators. SDT dealing with market risk, compliance with norms, and support to cooperatives should include provisions to avoid their leading to price controls and other forms of restrictions to operations by private traders.

At a more general level, SDT implies delayed reform by African and other developing countries. As shown by most studies, a significant share of the potential benefits from global trade reform would come from changes in policies in developing countries. Anderson et al. (2005), for instance, conclude that "reform by developing countries is nearly as important in terms of economic welfare gains to the South as reform by high-income countries". Furthermore, if liberalization were to follow the tiered formula proposal in the current modalities, developed countries would reap 90% of the gains from reforms. If that is the case, then the cost and risks of SDT would significantly reduce its value to African and other developing countries.

KEY ELEMENTS OF SUCCESSFUL DOHA OUTCOMES FOR AFRICAN COUNTRIES

One often reads in the literature that reform of policies in OECD countries is politically unfeasible. Yet, OECD leaders are at the forefront of trade negotiations efforts. Either they believe that changes are possible or they are convinced that the current situation is increasingly politically unacceptable and thus have to display a willingness to act. Whatever the case, African countries cannot and should not buy into that argument. They have the most to lose under a continuation of global protectionism. Their efforts to achieve sustainable growth would be significantly hampered. On the other hand, OECD countries do have choices. They have the possibility to choose instruments that help them achieve their goals in the agricultural sector while not harming African economies. The difficulty for African countries is that global negotiations are based on a philosophy that places the emphasis on give and take, a mutual removal of harms caused by economic policies. If one party is hurt by the policies of another party but the former is not in a position to remove some harm that is caused by its own economic policies on the latter – because such harm is limited or does not exist, such as in the case of African countries – then it becomes difficult, if not impossible, for the former party to obtain satisfaction.

So far, trade negotiations have been effective in dealing with mutual harms. They have not been able to deal with a situation where the harm is in one direction and the economy that is being harmed has no economic means of pressure on the perpetrating country. African countries find themselves in this situation with respect to global protectionism. Considerable pressure has been exerted on developed countries at the beginning of the Doha Round to shift the negotiating philosophy to deal with unidirectional or absolute harm, when the policies of a given country are causing considerable harm to another, which is not in a position either to retaliate or to offer some type of economic reprieve. The strategy here has been to move towards a so-called development round. Looked at carefully, the rationale underlying the 'development round' concept is one that is based on the unlikelihood or impossibility of African policies to cause harm. The underlying argument is that they have not caused harm now, and hence do not have to be targeted for reform in the negotiations; they cannot cause harm in the future, and thus should be exempted from future agreements through SDT and other types of derogation.

As pointed out earlier, the consequence of this strategy is to legitimize the perpetuation of global protectionism from the point of view of African countries and their advocates. It has, however, been quite helpful in taking the negotiations away from the offer and counter-offer paradigm. In order to be really helpful to African countries, the strategy should be expanded to recognize the right of African economies to equal opportunity to compete. Two decades of bilateral and multilateral conditionalities and reforms to rid their economies of policy distortions give them the right to expect removal of distortions and policy interventions, in particular, in the countries that have supported and helped enforce these conditionalities. African countries also have the right to expect the discourse about globalization and their integration into the world economy to be reflected in the rules and principles governing global trade and economic relationships between countries. A basic principle is that economic relationships and exchange between countries be based on the market mechanism. African countries have the right to demand that these principles be also extended and fully applied to agricultural trade. Whether or not they succeed in obtaining satisfaction during the Doha or subsequent rounds should not change this position. More importantly, they should have no interest in entering into agreements that would keep them away rather moving them closer to that outcome.

Furthermore, the efforts by OECD countries and multilateral organizations to mobilize the world in eliminating poverty in African and other developing countries should dictate greater efforts towards effective liberalization of global agricultural markets. A continuation of global protectionism would starkly reduce the capacity of many African countries to achieve faster and broad-based growth. It would also reduce returns to official development assistance (ODA) and other efforts to spur growth in these countries. Vast segments of the population in these countries would continue to suffer the vagaries of international markets resulting from global protectionism.

Figure 4 illustrates the vulnerability of the poorest segments of the population in Africa to developments in international agricultural markets. Contrary to what many may think, many poor households in Africa depend on export crops for their livelihood, as indicated by the share of cotton in the incomes of poor rural households. In the present case, the poorest 40% among rural households derive about 20% of their income from cotton against less than 15% among the richest 20% of households. The right-hand-side graph shows the impact of falling world-market prices on poverty among rural households. A one percent decline in the world market price of cotton translates to a 0.5% decline in average incomes. It raises the poverty incidence (P0), or the number of households below the poverty line, by 1.5%. It increases the poverty gap (P1), or the difference between the average income of poor households and the poverty line, by nearly 2%. It makes the poorest among the poorest poorer by increasing the poverty depth (P2) by 3.5%.

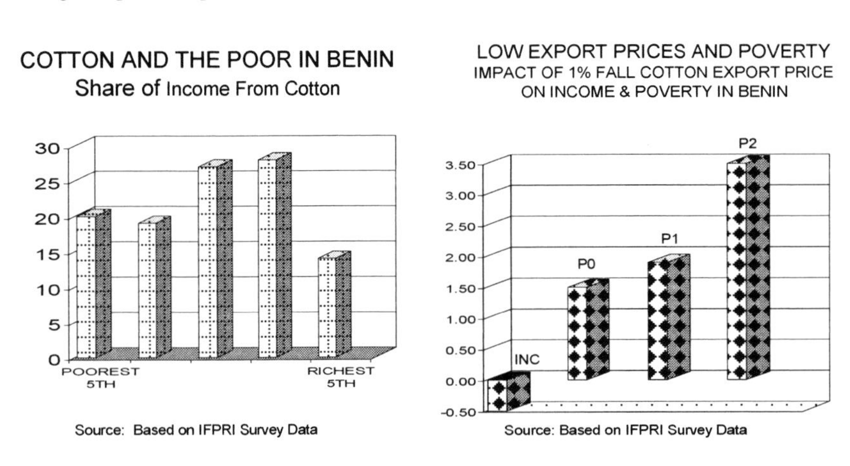

***Figure 4**. World markets and poverty in Africa*

Based on the arguments laid out above, successful negotiations from the point of view of Africa's long-term development interest should include: (i) effective decoupling of domestic support measures; (ii) full elimination of export subsidies; (iii) removal of border protection, including in emerging and middle-income countries; (iii) pursuit of the reform agenda in African countries; and (iv) disciplined SDT targeted as much as possible to compliance assistance. Preferences and untargeted SDT would just serve to legitimize global protectionism and hinder global liberalization efforts. Even, in this case, it is unlikely that agreements would be reached for all developed counties to provide EBA-style duty and quota free access for imports from African countries, as is timidly proposed in the current modalities. Moreover, such arrangements would need to be made binding to some extent to make some difference to long-term investment and growth. Such an option is currently not on the table. Rather, countries are being asked to consider EBA-style preferences on a voluntary and autonomous basis. Still, EBA-style preferences by developed countries would not open access to the faster growing markets in emerging countries to African exports. As illustrated in Bouët et al. (2001), market access measured by MAcMaps' aggregate measure of protection, which converts and sums up the *ad valorem* equivalents of various instruments of protection, is also highly restricted among emerging and other developing economies. The numbers that are reported for a sample of developing countries including Brazil, China and Morocco show overall food and agricultural protection and tariff peak levels that are similar to or higher than the levels observed in the US, EU or Japan. The ranking of countries by degrees of overall protection places these countries ahead of the US, EU and Japan.

Africa would still need some type of SDT in the case of full liberalization, but it would focus on compliance assistance rather than preferences and derogation. If SDT measures should involve derogation, they would have to be disciplined and rules defined for their applicability in cases where there is substantial risk of abuse. African governments would have to improve governance and economic management significantly. They would have to invest in business skills development, quality management systems, research and infrastructure. Although they will need assistance for a while to come, there is quite a bit that African countries can do on their own in these areas. For instance, the types of 'development interventions' that Dorward et al. discuss in their chapter in this volume could apply here, but not as substitutes to global trade liberalization. Rather, they could be considered accompanying measures. All three types of interventions – supply chain coordination, pump priming of investments, and threshold shifting – can be carried out through public partnership programs.

More problematic would be proposals for African countries to resort to protection, as implied partly under the threshold-shifting intervention option. Although one can try to make a historic case for protection, as in the chapter by Koning, there is enough evidence in the literature to show that the price instrument is too costly in terms of its intersectoral and economy-wide ramifications. Protection may have worked historically, perhaps, because it was an answer to slow and long-term changes in comparative advantages. It would be a poor answer to relative cost changes that result from foreign production and export subsidies that can be varied at will and overnight to wipe out any benefit from protection by African countries. Moreover, protection in African countries in the context of continued global protectionism would fail for the simple reason that African countries could not possibly 'outprotect' OECD countries. And even if they could, the cost and level of required protection among African countries would be lower with lower levels of global protectionism. Protection by African countries can therefore not be seen as an alternative to global trade liberalization.

Also, the research community will have to be significantly more relevant and helpful to African governments as they strive to cope with the effects of globalization and international protectionism. Rather than investigating Africa-wide implications of global trade liberalization, which is helpful in highlighting the overall cost of protectionism, or assuming away the capacity of African countries to adjust positively to changes in the global trading environment in the case of liberalization, the research community could be looking at investment and policy options that would help individual African governments craft strategies to gain from such liberalization. There is no Africa-wide government that can apply lessons and recommendations that are drawn from studies where African countries are lumped together in an artificial construct. Also, research that inherently assumes that African governments would respond to liberalization with little or no changes in strategies should be neither helpful nor relevant. We know the implications of the *status quo* and they are disastrous for African development, as indicated by the preceding discussion. The real issue is how to make effective liberalization beneficial for African countries, a question that can only be answered by country-level research that targets necessary investment and policy adjustment options.

CONCLUDING REMARKS

African countries are more vulnerable to global protectionism in agriculture than any other region of the world, due to the key characteristics of their economies. They have been affected negatively by the combination of domestic subsidies, border protection, unfair export competition and distortions in global markets. The preferences that have been introduced to mitigate some of the effects of global protectionism on African counties turn out to be less beneficial than hade been hoped for. In the aftermath of the Uruguay Round and in particular during the Doha Round, African countries and their supporters have placed substantial emphasis on Special Differential Treatment and other derogation to global trading agreements. While the value of preferences has been shown to be limited, SDT can involve significant risks and be open to abuse. Perhaps African countries have given up the hope that global protectionism can be reformed. Ironically, by positioning themselves for preferences and SDT, they help perpetuate and legitimize international protectionist policies.

Most studies conclude that African countries would gain from effective global trade liberalization. These gains are higher when long-term productivity adjustments are taken into consideration. By failing to reform international protectionist policies effectively, global trade negotiations have not responded to the real needs of the African economies. The philosophy of offer and counter-offer characterizing these negotiations is not geared towards addressing situations where there is absolute and not relative or mutual harm, in other words, a situation where the harm is unidirectional and the party being hurt is not in a position to offer some type of economic reprieve as an incentive for concessions from the party perpetrating the harm. It is only when negotiations are ready to deal with absolute harm that they will be able to address the case of African countries effectively and satisfactorily. The introduction of food security and development objectives into the Doha Round is a step in the right direction. The next step would be to reduce the emphasis on preferences and untargeted and undisciplined SDT in seeking to address these objectives under the framework of global trade agreements.

Elimination of the detrimental impact of global protectionism on economic growth and development prospects among African countries would require effective liberalization of global agricultural policies. This should be the objective of African countries under international trade negotiations. Any SDT to be sought and granted under such negotiations should be targeted to compliance facilitation and disciplined in order to avoid possible abuse. The time for such dramatic changes may not have arrived yet. The Doha agenda and the current modalities are too timid to lead to any significant reduction in actual support, subsidies and access barriers during this round. African countries will have to look beyond Doha, accept the reality of another lost decade, and use the coming years to prepare themselves appropriately for maximum success during the next round.

NOTES

[1] See Johnston and Mellor (1961), Lewis (1954), Michaely (1977), Heller and Porter (1978) and Balassa (1978). More recent work on the link between trade, economic growth and poverty can be found in Dollar and Kraay (2004), Winters (2002; 2004) and Winters et al. (2004).

[2] Such demands often target the more distorting and rent-inducing elements of OECD policies than the more justifiable interventions such as investments in infrastructure and research, for instance.

[3] For recent estimates, see Van de Mensbrugge and Beghin (2005), Anderson (2003; 2004), Francois et al. (2003), William Cline (2004), Ianchovichina et al. (2001) and Hoekman et al. (2001).

[4] See Anderson et al. (2005), Tables 16, 12.3, 12.12 and 13.

[5] The modalities and disciplines cited in this section are described in WTO (2003; 2004).

[6] Simulation results in Anderson et al (2005) suggest that application of 'sensitive product' status to 2% and 4% of production in developed and developing countries, respectively, would reduce gains from reforms by about 80%. Furthermore, protection measures based on CEPII's MAcMaps tariff measures reported in Bouët et al. (2001) indicate that elimination of tariff peaks alone would significantly reduce the aggregate rate of tariff on agricultural and food products.

REFERENCES

Anderson, K., 2003. *Trade liberalization, agriculture, and poverty in low-income countries.* UNU/WIDER, Helsinki. Discusssion Paper UNU/WIDER no. 2003/25. [http://www.wider.unu.edu/publications/dps/dps2003/dp2003-25.pdf]

Anderson, K., 2004. Subsidies and trade barriers: Copenhagen consensus challenge paper. *In:* Lomborg, B. ed. *Global rises, global solutions.* Cambridge University Press, Cambridge. [http://www.copenhagenconsensus.com/Files/Filer/CC/Papers/Subsidies_and_Trade_Barriers_140504.pdf]

Anderson, K., Martin, K. and Van der Mensbrugghe, D., 2005. Market and welfare implications of Doha reform scenarios. *In:* Martin, W. and Anderson, K. eds. *Agricultural trade reform and the Doha development agenda.* Oxford University Press, Oxford. [http://siteresources.worldbank.org/INTRANETTRADE/Resources/239054-1109114763805/Ch12_AndersonMartinMensbrugghe.pdf]

Baffes, J. and De Gorter, H., 2004. *Disciplining agricultural support through decoupling.* Mimeo no. p. 1.

Balassa, B., 1978. Exports and economic growth: further evidence. *Journal of Development Economics,* 5 (2), 181-189.

Bouët, A., Fontagné, L., Mimouni, M., et al., 2001. *Market access maps: a bilateral and disaggregated measure of market access.* CEPII, Paris. CEPII Document de Travail no. 18. [http://www.cepii.fr/anglaisgraph/workpap/pdf/2001/wp01-18.pdf]

Brenton, P. and Ikezuki, T., 2005. The impact of agricultural trade preferences, with particular attention to the least developed countries. *In:* Aksoy, M.A. and Beghin, J.C. eds. *Global agricultural trade and developing countries.* World Bank, Washington.

Cline, W., 2004. Industrial-country protection and impact of trade liberalization on global poverty. *In:* Cline, W. ed. *Trade policy and global poverty.* International Institute of Economics, Washington, 105-168. [http://www.iie.com/publications/chapters_preview/379/3iie3659.pdf]

Delgado, C., Hazell, P., Hopkins, J., et al., 1994. Promoting intersectoral growth linkages in rural Africa through agricultural technology and policy reform. *American Journal of Agricultural Economics,* 76 (5), 1166-1171.

Dollar, D. and Kraay, A., 2004. Trade, growth, and poverty. *Economic Journal,* 114 (493), 22-49.

Francois, J., Van Meijl, H. and Van Tongeren, F., 2003. *Economic implications of trade liberalization under the Doha Round: paper for 6th Annual conference on global analyis, The Hague, 12-14 June 2003.*

Heller, P.S. and Porter, R.C., 1978. Exports and growth: an empirical re-investigation. *Journal of Development Economics,* 5 (2), 191-193.

Hoekman, B., Ng, F. and Olarreaga, M., 2001. Tariff peaks in the quad and least developed country exports. *In:* Hoekman, B., Mattoo, A. and English, P. eds. *Development, trade, and the WTO.* World Bank, Washington.

Ianchovichina, E., Mattoo, A. and Olarreaga, M., 2001. *Unrestricted market access for Sub-Saharan Africa: how much is it worth and who pays?* Centre for Economic Policy Research, London. Centre for Economic Policy Research Discussion Paper no. 2820.

Jaffee, S.M. and Henson, S., 2005. Agro-food exports from developing countries: the challenges posed by standards. *In:* Aksoy, A. and Beghin, J.C. eds. *Global agricultural trade and developing countries.* World Bank, Washington.

Johnston, B.F. and Mellor, J.W., 1961. The role of agriculture in economic development. *The American Economic Review,* 51 (4), 566-593.

Lewis, W.A., 1954. Economic development with unlimited supply of labour. *The Manchester School,* 22 (2), 139-191.

Liedholm, C., McPherson, M. and Chuta, E., 1994. Small enterprise employment growth in rural Africa. *American Journal of Agricultural Economics,* 76 (5), 1177-1182.

Michaely, M., 1977. Exports and growth: an empirical investigation. *Journal of Development Economics,* 4 (1), 49-53.

Oyejide, T.A., 2002. Special different treatment. *In:* Hoekman, B., Mattoo, A. and English, P. eds. *Development, trade, and the WTO.* World Bank, Washington.

Urata, S., 1989. Sources of economic growth and structural change: an international comparison. *In:* Williamson, J.G. and Panchamukhi, V.R. eds. *The balance between industry and agriculture in economic development. 2, Sector proportions.* MacMillan Press, London, 144-166.

Van der Mensbrugghe, D. and Beghin, J.C., 2005. Global agricultural reforms: what is at stake. *In:* Aksoy, A. and Beghin, J.C. eds. *Global agricultural trade and developing countries.* World Bank, Washington.

Wilson, S.W., 2002. Standards, regulations, and trade: WTO rules and developing country concerns. *In:* Hoekman, B., Mattoo, A. and English, P. eds. *Development, trade, and the WTO.* World Bank, Washington.

Winters, L.A., 2002. Trade liberalisation and poverty: what are the links? *The World Economy,* 25 (9), 1339-1367.

Winters, L.A., 2004. Trade liberalization and economic performance: an overview. *Economic Journal,* 114 (493), 4-21.

Winters, L.A., McCulloch, N. and McKay, A., 2004. Trade liberalization and poverty: the empirical evidence. *Journal of Economic Literature,* 62 (1), 72-115.

WTO, 2003. *Negotiations on agriculture: first draft of modalities for the further commitments.* WTO, Geneva. [http://www.wto.org/English/tratop_e/agric_e/negoti_mod1stdraft_e.pdf]

WTO, 2004. *Doha work programme: decision adopted by the General Council on 1 August 2004.* WTO, Geneva. [http://trade.ec.europa.eu/doclib/docs/2004/august/tradoc_118356.pdf]

CHAPTER 10

THE PRACTICAL EXPERIENCE WITH AGRICULTURAL TRADE LIBERALIZATION IN ASIA

DAVID DAWE

Senior Food Systems Economist, Food and Agriculture Organization (FAO), Regional Office for Asia and the Pacific[1], *Bangkok, Thailand*

INTRODUCTION

Asia is a large continent, home to most of the world's poor, and agriculture is the largest employer of the poor. The effects of agricultural trade liberalization in Asia are thus very important. Unfortunately, as best as I can tell, the literature on ex-post assessments of agricultural trade liberalization in Asia is somewhat thin. Much of the economics literature concentrates on ex-ante assessments of trade liberalization, often with the help of computable general equilibrium (CGE) models. CGE models are an important avenue of research, because in principle they allow us to take account of the effects of agricultural trade policies on labour markets, foreign exchange markets, the manufacturing and service sectors, and other parts of the economy, even if in practice the analysis is often imprecise due to the large quantities of data that are required (and often not available). Yet ex-ante assessments of future trade liberalization are not enough – we need to be informed by past experience with trade liberalization as well as by models.

In principle, it is possible to conduct large-scale econometric studies of the ex-post effects of trade liberalization. Indeed, there is a large literature on trade/openness and growth that argues that some measures are correlated with growth, e.g. trade as a percentage of GDP, or levels of tariff and non-tariff barriers. But any such study will run into difficulties because, at least at the commodity level, these measures are not necessarily correlated with trade liberalization. For example, imports as a percentage of domestic consumption (or exports as a percentage of domestic production) could decline at the same time that tariffs decline and world prices and domestic prices move closer together. As an example, the ratio of rice

N. Koning and P. Pinstrup-Andersen (eds.), Agricultural Trade Liberalization and the Least Developed Countries, 175–195.

imports to consumption declined for the Philippines and Sri Lanka in the 1970s as the new Green Revolution technology increased rice productivity in these traditional importers, leading to a convergence of world and domestic prices. Thus, a decline in protection could be consistent with a decline (rather than an increase) in trade.

Even measures of tariff and non-tariff barriers (NTB) are not necessarily correlated with trade liberalization. For example, if a tariff declines from one prohibitive level to another prohibitive level, this is not really trade liberalization. On the other hand, trade liberalization may occur without any legal changes in NTB if the licensing procedure for obtaining access to imports becomes more transparent and liberal. At a more aggregate level, it is well-known that average tariffs can be misleading because this procedure gives equal weights to minor commodities that may not be important to trade. Trade-weighted tariffs solve this problem but create another, namely that high tariffs can lead to minimal trade and low or zero trade weights. Thus, while it is trivial to measure trade liberalization in an ex-ante sense (an analyst or modeller can define any policy change (s)he likes), it is more complicated in an ex-post sense, especially at an aggregate level. It is usually possible to define the timing of major liberalization events in a specific country without too much difficulty, but it is much harder to quantify the magnitude of the liberalization that has occurred, which of course is an essential precondition to a cross-country econometric study.

Other measures of trade liberalization are of course possible and useful, including data on the extent of subsidies and various measures of domestic support. Export subsidies are not at all common in Asia, however, so that is not a useful definition for the present paper. In developing Asia, domestic support is also much less important than in OECD countries.

Thus, in trying to understand the ex-post impact of trade liberalization it may be more productive to take a case study approach, focusing on specific commodities in particular countries at particular times. This is a more difficult research agenda, because many case studies are necessary in order to reach general conclusions. Although the costs of this research are large, I believe that the potential gains are also large, including an increased opportunity for productive dialogue between those who are generally supportive of trade liberalization and those who are quite sceptical.

The general objective of this paper is to make an ex-post assessment of the agricultural trade liberalization that has occurred to date in Asia. In order to meet this objective, the paper has three main sections. First, given the caveats mentioned above, the paper will try to make a broad assessment of how much liberalization has occurred by looking at trends in tariffs, NTB and trade, including the role of the WTO in this process. Second, the paper will examine a series of case studies for useful insights on how agricultural trade liberalization has worked in actual practice, although I do not have the data or literature references required to elaborate on a large number of case studies. No attempt will be made to quantify the effects of agricultural trade liberalization on the overall growth process. The final section of the paper will conclude with a discussion of agricultural protectionism, equity and growth in the Asian context, a discussion of circumstances when agricultural protection may be justified, and a suggestion that future research needs to examine more carefully the effects of trade liberalization on the losers who are inevitably created.

HOW MUCH HAS AGRICULTURAL TRADE BEEN LIBERALIZED IN ASIA?

As noted above, it is often difficult to decide whether or not trade liberalization has occurred. With that caveat in mind, many countries in Asia have lowered tariffs on agricultural products substantially during the past 25 years. For example, the average agricultural tariff in the Philippines was lowered steadily from 62% in 1980 to just 10% by 2003 (*Trade liberalization, agriculture and small households in the Philippines: proactive responses to the threats and opportunities of globalization* 2004). In Bangladesh, from the late 1980s to the late 1990s, applied tariffs declined for nearly all agricultural product categories (e.g. the applied tariff on edible oils declined from a range of 50-100% to 30-60%; FAO 2000). Another study on Bangladesh found that the unweighted average tariff in agriculture declined from 76% in 1991-92 to 34% in 1998-99 (Dowlah 2003). A similar trend, from 34% to 12%, was noted for import-weighted tariffs. Dowlah (2003) also found that the effective rate of protection in agriculture declined from 70% in 1992-93 to 21% in 1999-00. In Pakistan, the maximum applied tariff rate of ordinary tariffs on agricultural products declined from 225% in the late 1980s to 35% by the late 1990s (FAO 2000). China lowered its most-favoured nation (MFN) tariffs on agricultural products from 46% in 1992 to 19% in 2001, while India lowered its MFN tariffs from 66% in 1990 to 42% in 2001 (FAO 2005).

Non-tariff barriers to trade in agricultural products have also been declining. In India, import controls on sugar and cotton were removed in 1994, and in 1995 almost all edible oils were placed under the Open General License System (FAO 2000). Exports of ordinary rice were allowed beginning in 1994. In Bangladesh, only three Harmonized System (HS) lines had quantitative restrictions remaining by 1994, compared to much more extensive restrictions in earlier years. Import procedures have also been deregulated (FAO 2000). In Pakistan, import quotas were steadily eliminated between 1987 and 1995 until few remained (FAO 2000). In Sri Lanka, quantitative restrictions have been removed on all agricultural products except wheat and wheat flour (Dorosh 2003).

Regional and bilateral free-trade agreements are also becoming important. The Association of Southeast Asian Nations (ASEAN) Free Trade Area (AFTA) has undertaken to eliminate all NTB within agriculture by 2010, even for sensitive products such as rice, and to keep tariffs at very low levels. ASEAN is negotiating with China, Japan and Korea for a wider free trade area, and the Southeast Asian countries are also negotiating bilaterally with these same countries. For example, Thailand and China have concluded a free trade agreement on vegetables. The Philippines and Japan are also in the midst of free trade negotiations. The member countries of the South Asian Association for Regional Cooperation (SAARC) have also negotiated a South Asian Free Trade Agreement (SAFTA). Indeed, there are so many negotiations that it is next to impossible to keep track of them all, although bilaterals.org (2006) provides an excellent up-to-date summary.

However, trade is not becoming more liberal for all commodities in all countries. For example, in the early 1980s, domestic rice prices in the Philippines were roughly in line with world market prices. Since 1985, however, domestic wholesale prices have on average been 66% above import parity prices. The discrepancy has widened in recent years, to 89% on average since 1993. After allowing private sector rice imports starting in 1999, Indonesia has recently forbidden all rice imports for certain months of the year, and instituted new licensing procedures for rice imports during other months. In Sri Lanka, tariffs on chilies, onions, pulses and maize increased from 5% in 1986-88 to 35% by 1999 (FAO 2000). These instances of increasing protection are driven by different factors – falling world rice prices in the case of the Philippines (coupled with a desire to keep domestic prices roughly stable in real terms), new-found lobbying power by rice producers in Indonesia with the advent of democracy, and sharp reductions in area planted to onions in Sri Lanka (before the increase in tariffs restored the area to its earlier levels).

In addition to the general lowering of tariff and non-tariff barriers, it also appears that trade in agricultural products has generally expanded more rapidly than production and consumption[2]. This is especially true in Japan and Korea (see Table 1), where the ratio of imports to consumption has been increasing steadily and substantially for the past 40 years for a number of foods: fruit, pulses, rice, roots and tubers, vegetables, vegetable oils and wheat. The share of imports in consumption for maize has also increased, albeit much more slowly, because even in the early 1960s import dependence was already very high. Not surprisingly, the share of exports in production has not increased substantially for these countries, whose comparative advantage does not lie in agriculture. Increased agricultural imports by Japan and Korea do not necessarily indicate any changes in trade policies – most of the increase is due to rising incomes coupled with a lack of comparative advantage in agriculture due to land scarcity. Nevertheless, more imports do indicate a willingness to participate more in international trade – trade barriers could have been erected to staunch this flow. Indeed, such policies are in place to dramatically slow the opening of the rice sector in these economies.

Southeast Asia's agricultural trade has also expanded considerably in many food products. The main exception has been maize exports, which have declined because of sharp increases in demand for feed. In South Asia, trade has expanded particularly rapidly for vegetable oils, as imports have surged both in India and the rest of South Asia. Trade has contracted substantially in the case of wheat imports, but this is due to the Green Revolution that transformed the region into approximate self-sufficiency after being a large net importer in the early 1960s. Thus, even where trade has declined, the explanation rests with factors other than a retreat from liberal trade.

Table 1. *Change in trade importance (percentage points), various commodity groups and countries/regions, between 1961-1965 and 1998-2002*

Exports	India	Other South Asia	Southeast Asia	China	Japan and Korea
Fruits	0.00	0.03	0.10	-0.08	-0.04
Maize	0.00	0.00	-0.14	0.06	0.00
Pulses	0.01	0.03	0.11	0.13	-0.01
Rice	0.06	0.01	0.00	0.01	0.01
Roots and tubers	-0.01	0.01	0.26	0.00	0.00
Sugar	-0.09	0.09	-0.07	-0.58	0.11
Vegetable oils	-0.01	-0.27	0.38	0.01	-0.02
Vegetables	0.00	0.01	0.03	0.01	0.00
Wheat	0.02	0.02	0.03	0.00	0.03
Imports					
Fruits	0.00	0.02	0.03	-0.03	0.33
Maize	-0.02	0.08	0.13	0.02	0.05
Pulses	0.08	0.27	0.03	0.04	0.31
Rice	-0.03	-0.01	-0.02	-0.01	0.02
Roots and tubers	0.00	-0.02	0.03	0.03	0.36
Sugar	0.03	-0.26	0.17	-0.43	0.08
Vegetable oils	0.40	0.46	0.00	0.20	0.30
Vegetables	0.00	0.01	0.04	0.00	0.11
Wheat	-0.27	-0.12	0.05	-0.21	0.26

Trade importance is defined as exports or imports (in quantity terms) as a share of either production (for net exporters) or consumption (for net importers), the latter also in quantity terms. Data on the change in trade importance are in percentage points, not percent.
Source of raw data: FAO (2006).
Other South Asia includes Pakistan, Bangladesh, Sri Lanka, Nepal, Bhutan, and Maldives.
Southeast Asia includes Indonesia, Philippines, Viet Nam, Thailand, Malaysia, Myanmar, Cambodia, Laos, Brunei, and Timor-Leste.
Japan & Korea excludes North Korea.

Shaded cells indicate those commodities and countries/regions for which trade was less important from 1998 to 2002 compared with 1961 to 1965.

The general lowering of tariffs and NTB, coupled with widespread growth in the ratios of imports and exports to consumption/production, suggests that agricultural trade in Asia today is more liberal than it was 10 to 15 years ago. However, generally speaking, the legal conditions imposed by the Uruguay Round Agreement on Agriculture (AoA) have not been responsible for this outcome. Most Asian countries currently have applied agricultural tariffs that are much lower than the legal bindings mandated by the WTO (Ingco 1995; FAO 2000; 2005). There are, of course, exceptions. For example, India has bound its tariff on soybean oil at 45%, which has effectively prevented it from raising tariffs on competing vegetable oils (e.g. palm oil) to even higher levels, as high tariffs on palm oil would just lead to more imports of soybean oil (Landes and Gulati 2004). The Uruguay Round also forced Japan and Korea to open their domestic rice markets, and this has led to more imports by these countries. (On the other hand, Japan has also increased its exports of rice in recent years, reducing the effect of the AoA on its net trade position.) Generally, however, there are not many of these examples. In other words, Asia has liberalized voluntarily – it has not been dragged kicking and screaming into the process.

THE EFFECTS OF AGRICULTURAL TRADE LIBERALIZATION: SOME CASE STUDIES

It is perhaps helpful to distinguish different types of trade liberalization. Because agriculture is sometimes protected and sometimes taxed, agricultural trade liberalization can either remove protection or remove taxation. For those who argue that agricultural growth is crucial for poverty alleviation and kick-starting growth in poor economies, the latter would be expected to be beneficial trade liberalization, while the former might be expected to have negative effects. Included here in the category of 'beneficial' (in the sense of providing larger incentives for agricultural production) trade liberalization would be removal of protection for agricultural input industries such as irrigation, fertilizers and seeds.

Some examples of trade liberalization that improved incentives for agriculture are the reform experiences of China and Viet Nam. Clearly the opening of agricultural markets in China beginning with the household responsibility system in 1978, and the *doi moi* experience in Viet Nam, had major beneficial effects for the agricultural sectors and economies of these two countries. However, most countries today do not have the opportunity to realize such gains because their agricultural economies are not as rigidly controlled as those of China before 1978 and Viet Nam before 1986, so these reforms are not discussed further[3]. Other examples of 'agriculture-promoting' trade liberalization include the irrigation and fertilizer sector reforms in Bangladesh and the seed industry reforms in India. These reforms are discussed below in more detail. Not all of the case studies below describe trade liberalization episodes – some simply describe how changes in price incentives (not necessarily due to trade liberalization) have affected the given sector. These experiences are included because if trade liberalization is believed to affect farmers through changes in prices, then any change in price might give us useful insights, not just changes in prices driven by changes in trade policy.

Before proceeding to the case studies, I will report the results of one crude macro type of calculation. One fear of liberalization is that it will throw many farmers out of business, or at minimum, force them to switch crops, as happened when Sri Lanka liberalized trade in potatoes and onions and when India liberalized trade in vegetable oils. However, the increased liberalization that appears to have occurred (as discussed in the previous section) does not appear to have substantially increased the frequency of large reductions in area for specific crops. A calculation of the percentage of commodities undergoing a 10% reduction in area (comparing two adjacent three-year periods) for several countries in Asia, individually or as a group, shows no discernible trend in the frequency of such events during the past 40 years (Figure 1). This outcome is consistent with a relatively smooth transition to increased trade. It is also consistent with countries resisting liberalization in the commodities where the largest displacements might be expected to occur. This is admittedly a crude measure, but it does appear to indicate that abrupt transitions for farmers out of specific crops are not increasing over time.

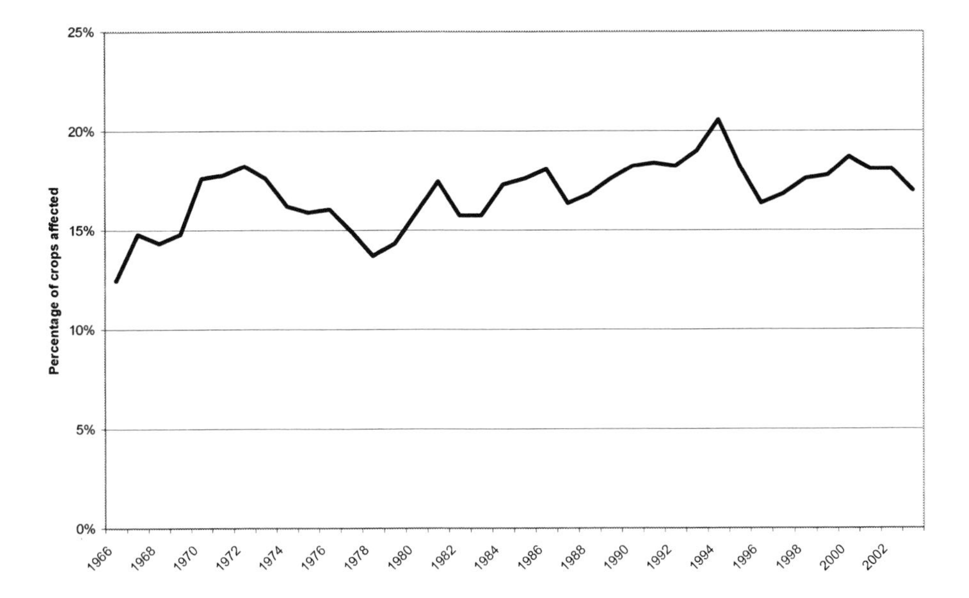

Source of raw data: FAO (2006).

***Figure 1**. Incidence of negative area fluctations greater than 10%*

Rice price policy and productivity growth in Southeast Asia

It is important to realize that, when it comes to rice price policy in Asia, protection or taxation (a positive or negative nominal protection coefficient, NPC) is rarely a conscious, active policy choice. Rather, the level of the NPC is often the outcome of a desire for long-term stability in domestic prices (in real terms), coupled with exogenous changes in world rice prices due to either changes in the US dollar price or changes in the exchange rate. In other words, the NPC is determined primarily by the exchange rate or the world market dollar price, both of which are exogenous to agricultural policymakers (Timmer 1993). For example, most of the increased protection for rice in Japan in the 1970s and 1980s was due to appreciation of the yen and declines in the US$-denominated world rice price, not because of higher real yen prices received by Japanese farmers. Changes in rice price policy in Asian countries tend to be passive, not active.

Thailand. Whatever the causes, the experience with rice price policy in Southeast Asia has been diverse. Thailand followed a policy of keeping domestic prices well below world prices for many years. By 1986, policy had changed to one of free trade (trade liberalization), with domestic prices subsequently tracking world prices almost exactly. This transition from a negative NPC for rice to free trade was facilitated by the sharp decline in the baht equivalent of world prices from 1981 to 1986, which eliminated the need for an increase in the level of domestic prices in order to remove taxation of farmers. While yields had been increasing before the removal of taxation of farmers, yields did begin to rise more rapidly after 1990. However, it is unlikely that removal of farm taxation was the cause for this increase, as domestic rice prices fell during the 1990s due to declining world prices (i.e. farmers presumably respond to changes in prices, not changes in the NPC). Total factor productivity may have increased even more than yields due to the extensive mechanization that has occurred in recent years. But this mechanization was clearly a response to rising wages, not to increased levels of the NPC (from negative to zero).

In recent years, domestic wholesale prices have continued to track world prices, but government subsidies to farmers have pushed farmgate prices to levels that are above world price equivalents. This seems to be a political response to a widening income gap between urban and rural areas. Because of Thailand's higher level of per capita income and smaller proportion of farmers in the labour force (after many years of rapid growth), such subsidies are now more easily affordable by the government. Rice yields showed a small increase in the late 1990s, but this occurred before major increases in the support price program. Yields have been steady from 2000 to 2005, when the program was in full swing and stocks began to accumulate.

Indonesia. Indonesia followed a different course than Thailand. From 1969 to 1996, domestic rice prices were stabilized around the trend of world prices. In some years there was taxation, while in others there was protection, but these changes from taxation to protection and back were the result of changes in the rupiah equivalent of world prices, not changes in agricultural price policy (Timmer 1996). This long-term stabilization about the trend of world prices was facilitated by a sharp depreciation of the rupiah in real terms during the period, which kept the rupiah equivalent of world prices roughly constant in the face of a declining world price in dollar terms.

It is possible to view the exchange rate depreciations in 1978 and 1983 as a form of protection, although two points are worth noting in that regard. First, the depreciations can largely be viewed as a reaction to the 'Dutch Disease' phenomenon whereby greater oil exports increased the relative price of non-tradables (i.e., the real exchange rate). In other words, the depreciations removed the temporary artificial taxation of tradable goods. While this could be considered as tax-removing liberalization, the second point is that domestic farmgate rice prices were approximately constant in real rupiah terms during the period of the depreciations. In other words, the depreciations did not substantially improve the terms of trade for farmers, except relative to a hypothetical counterfactual where rice was a tradable good and there was no depreciation. But because the international rice trade was controlled by the government during this time, rice was not a tradable commodity (i.e. changes in world prices did not lead to changes in domestic prices). Thus, while the exchange rate depreciations increased the NPC, domestic farm prices in real terms were relatively constant.

Beginning in 1999, however, there was a sharp increase in protection and in the level of domestic farm prices in response to the rising political demands of rice farmers. Although rice farmers are not a majority of the Indonesian population, rice farming is the single most common occupation in the country, and they have been able to exert leverage on the political process. This is a clear counter-example to the general trend of trade liberalization. The increase in protection since 1999 does not seem to have generated any important productivity gains in the rice sector, although yields did increase slightly from 2001 to 2003. There is also no evidence from the field of any important productivity advances due to reduced input use. The dominant trend since 1990 has been a stagnation of rice yields. In contrast, productivity growth was rapid in earlier years when the NPC was zero on average and the real level of farm prices was constant, as new seeds from the Green Revolution and increased use of fertilizer in irrigated areas led to major yield gains and the temporary achievement of self-sufficiency in 1984.

Philippines. The Philippines' history of rice price policy is different yet again from that of Indonesia and Thailand (Dawe 2003). Domestic farm prices declined in real terms by approximately 40% from the mid-1970s to the early 1980s, as production gains from the Green Revolution led to increased supplies on domestic markets and eventually small quantities of exports. While prices declined steadily during the entire period, at no point did they appear to halt the continually increasing adoption of modern varieties and concomitant increases in yields. By the end of this period, domestic price levels were similar to world prices. From the early 1980s, however, the NPC for rice began to increase steadily, as world prices (in peso as well as dollar terms) fell sharply in the mid-1980s, and then once again in the late 1990s. In response to these developments on the world market, Philippine policy passively allowed very substantial levels of protection to develop (36% on average from 1985 to 1992, and 89% from 1993 to 2003) in order to maintain domestic farm prices roughly constant in real terms.

From 1991 to 1999, when protection became quite high, rice yields were stagnant, most likely reflecting nearly complete prior adoption of Green Revolution technology. Starting in 2000, there was a slight decline in real farm prices. Around the same time, yields appear to have started increasing to some extent, although no causation is asserted here.

Sri Lanka. In terms of nominal protection, Sri Lanka has followed a path similar to that of the Philippines, with high and increasing rates of protection since the early 1980s. The increasing protection coincided with a sustained decline in real domestic prices (i.e. domestic prices declined more slowly than world price equivalents, leading to an increase in protection). Despite the worsening incentives, rice yields were essentially stagnant during this period, except for a small increase toward the end of the 1990s. Thus, the decline in prices did not seem to cause any decline in yields. The stagnation in yields is consistent with the fact that modern varieties had been adopted on nearly 95% of Sri Lanka's rice area by 1980.

Based on this rudimentary analysis of rice in these four countries, it would seem that price policy in and of itself is unlikely to generate important gains in productivity: the availability of technology is crucial. Indeed, in the Philippines case, yields continued to increase substantially even as prices were falling, and something similar occurred in Thailand. Price policy may be able to play a supporting role in the presence of a new technology, especially if that technology is perceived as risky by farmers. Price policy could be supportive either by raising the average price through protection, or perhaps more important, by stabilizing the market price so that farmers do not have to bear this risk[4]. The productivity gains achieved through adoption of the new technology would then generate additional income in the rural sector that could be invested to increase productivity in other sectors, or spent on non-tradable goods produced in the rural sector, thus increasing demand and contributing to a 'big push' in rural areas à la Rosenstein-Rodan.

Liberalization of international rice trade in Asia

During the past 20 years, there has been a broad liberalization of the international rice trade in Asia and elsewhere (Dawe and Slayton 2004). The direct role of government in carrying out rice imports and exports has been reduced, with a decline in the importance of government to government contracts and an increased role for the private sector. Thailand effectively abandoned its export taxes by 1976, and Viet Nam has increased competition among state-owned enterprises as well as allowing more private sector participation in the export trade. Pakistan fully privatized rice exports in 1996, removing the monopoly formerly enjoyed by the Rice Export Corporation. Sri Lanka abolished the monopoly of the government parastatal on rice imports in 1990 and allowed private traders to import rice subject to a tariff (Dorosh 2003). During the same time, the proportion of world production that has entered international trade increased from about 4% prior to the mid-1990s to 6-7% today (more than a 50% increase in the size of the market). Trade liberalization by these exporters increased the stability of the world rice market, although technological factors (more irrigation, greater area planted to pest and disease resistant varieties) and economic growth (leading to falling per capita rice consumption and an increased supply of exports from Thailand) were also important in terms of improving price stability (Dawe 2002).

Against this backdrop of a more stable world rice market, Bangladesh liberalized rice imports in April 1994, allowing the private sector to import subject only to a tariff. India also liberalized its rice trade in late 1994, relaxing its ban on exports of ordinary rice and allowing more private sector participation (Del Ninno and Dorosh 2001). This pair of liberalizations paved the way for a surge of imports (about 2.25 million tons) by Bangladesh in 1998 in response to the 'flood of the century.' Most of these imports came in small shipments from India. Absent these imports, Dorosh (2001) estimates that rice prices in Bangladesh could have increased by 40 to 60%. Of course, if these imports had not arrived, the government would have stepped in to some extent, but Dorosh argues that "public sector imports of a magnitude equal to private sector flows would have been highly unlikely" due to funding constraints. In this case, trade liberalization that allowed domestic prices in Bangladesh to maintain parity with external prices, even in the short run, arguably made an important contribution to food security in Bangladesh.

In Indonesia, beginning in 1969, Bulog (the state logistics agency) had a monopoly on international rice trade, with decisions on import quantities being made ultimately by then-President Suharto (Timmer 1996). Beginning in 1999, however, Indonesia allowed the private sector to import rice subject only to a tariff of 430 Indonesian rupiah per kg (equivalent to about a 30% tariff at that time). This change was dictated by an agreement with the International Monetary Fund (IMF). Since that time, the private sector has been responsible for about three-fourths of Indonesia's imports (Dawe and Slayton 2004). During this time, there have been no major price surges or supply disruptions; participation of the private sector appears not to have compromised food security in the short run. On the other hand, the relatively high tariff meant that domestic prices began to diverge from international prices quite substantially, apparently in response to political lobbying by rice farmers in the newly democratic environment. The new, higher, rice prices were in contrast to a long history of stabilizing prices around the long-run international price. Thus, although the rice trade was liberalized in the sense of allowing more private sector participation, it became less liberal in terms of the convergence of domestic and international prices. This latter factor raised food prices and increased poverty.

More recently, however, the government instituted a temporary import ban (to coincide with the main harvest), and has also issued a decree that will allow only specially licensed importers to import rice (previously, only a general import license was required). It is unclear if tariffs will increase in the future or how restrictive the licensing procedures will be, and how these developments will affect the role played by private traders.

To summarize, it would seem that the experience with liberalization of international rice trade in Asia has been largely positive. Increased competition among exporters in Viet Nam has almost certainly benefited farmers, although it probably hurt some major state-owned enterprises. Private sector imports in Bangladesh made an important contribution to food security during the floods in 1998, and there were no major adverse impacts upon Indonesia's opening of the rice trade to the private sector. Part of the reason that the effects have been positive is the increased depth and stability of the world rice market, as well as the increase in the quantity and quality of transportation and communications infrastructure during the past 20 years (Rashid et al. 2005).

Oilseeds and edible oils in India

From 1961 until 1976, annual imports of edible oils in India were on average less than 70,000 tons. From 1977 to 1988, however, total imports of vegetable oils (primarily palm, soybean, rape and mustard oils) increased to an annual average of about 1.3 million tons, rising to 2 million tons by 1987. During this period, domestic area planted continued to increase, from 14.3 million hectares in 1977 to 17.4 million hectares by 1988. The increased imports and rising domestic production supported increased domestic consumption.

Beginning in 1988, imports, which were controlled by the state, began to decline sharply as the government attempted to boost production incentives and encourage adoption of new technologies (Landes and Gulati 2004). From 1989 to 1994, imports of vegetable oils fell to less than 400,000 tons per year. Domestic oil and oilseed prices increased, and by 1993, area planted to oilseeds had reached 25.1 million hectares, an all-time record. At least seven million hectares of cereal land was converted to oilseeds during this process. Yields of major oilseed crops also increased sharply during this time.

Policies changed in 1994, however, as the government opened imports to private traders subject only to tariffs that were well below bound rates. Imports of palm and soybean oil surged, and India became the world's largest importer of vegetable oils. Domestic prices fell, and per capita consumption increased (Landes and Gulati 2004). Not surprisingly, area planted fell to 20.7 million hectares in 2001. After several years of growth, yield levels of soybean, rape and peanuts either stagnated or declined. Between 1998 and 2001, area planted to peanuts, rapeseed and sunflower seed fell by a combined 4.4 million hectares. Total area planted to all crops fell by 6.3 million hectares during this time. While one cannot be sure what happened to the land formerly planted to oilseeds using only macro-data, these data strongly suggest that at least some of this land was not converted to other crops. It would be very interesting and helpful to know what happened to these farmers that went out of oilseed production when imports increased, to know if they were relatively wealthy or poor, and to understand what difficulties they faced in the adjustment process. On the other hand, it is also important to note that oilseed area at its recent trough in 2001 (20.7 million hectares) was still greater than it was before the closure of the domestic market in 1987 (when it was 17.4 million hectares). Thus, it seems likely that the farmers negatively affected by the relaxation of imports had only been planting oilseeds for a few years under the stimulus of government incentives. It would also be helpful to know if the decline in yields led to a decline in farmer income, or whether the lower yields were due to lower input use that left farm profits unchanged.

In this case, it appears that the effects of liberalization have been largely what economic theory would predict. Unfortunately, it is not possible to ascertain the equity effects of this liberalization without a detailed analysis of food consumption patterns by income class and the relative position of oilseed farmers in the income distribution. Liberalization may have adversely affected innovation in the oilseed sector, but on the other hand if liberalization had not occurred, resources would have remained in oilseeds and not moved to other sectors, presumably to the detriment of innovation in those other sectors. (In other words, protection is relative, and not all sectors can be protected simultaneously).

It seems unlikely that this liberalization created much growth in India's agricultural sector; rather, it was probably more a static reallocation of resources within agriculture (and in favour of non-agricultural sectors). To the extent that the liberalization lowered food prices, this may have increased the competitiveness of unskilled labour and led to more employment creation in services and industry, although I know of no studies on the magnitude of this effect.

Irrigation and fertilizer sector reforms in Bangladesh[5]

Before 1978, the Bangladesh Agricultural Development Corporation (BADC) monopolized all procurement, maintenance and installation of tube wells. By late 1988, after a process of gradual changes, the private sector could import and trade in pump sets for shallow tube wells (STW) at low import duties and without having to adhere to standardization requirements. As a result of these reforms, installation costs for STW fell substantially, and the area irrigated by tube wells increased from 0.67 million acres in 1981/82 to 4.90 million acres in 1991/92 (62% of irrigated land), an increase by a factor of 7 in a decade. The rate of increase was particularly rapid in the latter half of the 1980s after the elimination of standardization requirements. Further, irrigation water prices declined substantially after liberalization for small and large farmers alike.

The expansion of private sector irrigation led to an acceleration of growth in the area planted to high-yielding modern varieties. Modern varieties (MV) were first introduced to Bangladesh in 1966/67, but adoption was much slower than in other Asian countries due to the lack of irrigation. In the first decade after irrigation sector reforms, the area planted to MV in the dry (boro) season increased by 1.7% per year. This growth accelerated to 3.7% per year after the import liberalization in 1988, helped along by privatization of fertilizer distribution that made fertilizer more available in a timely fashion. The increased adoption of modern varieties led to a rapid increase in yields.

This discussion strongly suggests major beneficial effects of trade liberalization that led to more rapid growth in the rice sector, which accounts for about 75% of total planted area in Bangladesh. Undoubtedly the productivity growth spawned in the rice sector had many multiplier effects outside of rice and even outside of agriculture, although I am not aware of any studies that have attempted to analyse that issue. One possible negative effect of the liberalization would have been on the domestic machinery industry, which was presumably hurt by the increased imports of irrigation equipment. It seems unlikely that these negative effects could outweigh the positive effects, but it would be helpful to know more about how this sector was affected and whether it hurt indigenous innovation.

In one sense, the liberalization of trade in irrigation equipment might seem to have little to say about the effects of protection or taxation of agricultural commodities, because the liberalization did not occur for the agricultural commodity per se. However, the liberalization illustrates a general lesson in that protection of one sector can retard growth of other sectors that use the protected commodity as an input. Since agricultural products are used as inputs into other sectors, most commonly as a foodstuff to feed workers in industry and services, protection of primary agricultural products might retard growth of labour-intensive industries by raising wages to uncompetitive levels. The case of Bangladesh's liberalization of irrigation equipment and the technological change it spurred in a downstream industry (rice) at least gives one pause to consider whether trade liberalization for an important wage good such as rice (when it is heavily protected) might spur growth and technological change in downstream labour-intensive industries such as garments and electronics. Indeed, the garment industry is now a major source of employment in Bangladesh.

Hybrid seeds in India[6]

Traditionally, India has maintained restrictions on the size of companies allowed to participate in certain industries, in an effort to protect smaller firms. In 1986, however, the government removed such restrictions on companies in the seed industry. Two years later, restrictions were relaxed on imports of seeds and germ plasm for vegetables, coarse grains and oilseeds. As a result, private research expenditures increased from $1.2 million in 1987 to $4.7 million by 1995. Some of this new spending came from large firms, but some also came from small firms that were able to start viable small-scale research programs once imports of seed were allowed. The hybrid seeds developed by private firms for cotton, maize, sunflower, sorghum and pearl millet led to increased yields in many instances, or increased area, as for sunflowers in Karnataka. Thus, trade liberalization has led to increased productivity. The increased role of the private sector has probably increased competitive pressures on public sector plant breeding and may eventually lead to some layoffs there, but such outcomes are inevitable whenever competition is allowed.

DISCUSSION AND CONCLUSIONS

Agricultural protectionism, equity and growth in developing Asia

Price protection for agriculture (e.g. import restrictions on competing products, output price support, input subsidies) can be justified on efficiency grounds if there is a technology that is profitable at long-run prices but nevertheless may not be adopted by farmers who are risk-averse or do not have sufficient knowledge of the technology. In these circumstances, subsidies can spur farmers to take risks or acquire knowledge that will lead to adoption. But identification of such technologies is not necessarily easy, as there are always new technologies being promoted. Furthermore, tilting the terms of trade in favour of agriculture simultaneously tilts the terms of trade against other sectors, which then presumably discourages adoption of promising new technologies in those sectors.

In the absence of a new profitable technology, however, price protection for agriculture (or any other sector) simply redistributes income from one group to another. Such redistribution can have implications for both equity and growth. It obviously affects equity, but, at least in the Asian context, it is not quite so simple as redistributing income from relatively well-off urban dwellers to relatively poor rural residents. The complexity arises because of the large pools of functionally landless labourers in countries such as India, Indonesia, Bangladesh and the Philippines. These landless labourers are usually the poorest of the poor, do most of the production work, and are net consumers of food (Dawe 2004). For example, Mellor (1978) found that the two poorest deciles of the income distribution in India were net buyers of food, while each of the top eight deciles were net sellers of food. Sahn (1988) showed that 84% of rural households in Sri Lanka were net consumers of rice (and, of course, all urban households). The data he presented also show that two-thirds of farmers with marketable surpluses of rice are in the top half of the expenditure distribution. Balisacan (2000) estimated net rice consumption as a percentage of total consumption in the Philippines, and found that it was highest for the bottom two deciles of the expenditure distribution. For the bottom decile (decile 1), the share was approximately +7.5%, while for the second decile, it was about +2%. Net consumption was estimated to be negative (i.e., production exceeds consumption) for the middle of the distribution, that is, deciles 4 to 8. These data show that the poorest of the poor in the Philippines are net consumers of rice, not net producers. Thus, many poor people in rural areas would be hurt by higher food prices due to protection[7]. Unfortunately, nearly all of the popular trade liberalization debate that is carried out in a global context focuses on farmers, to the exclusion of the landless poor in rural areas.

Price policy could also have implications for economy-wide growth in the absence of a new technology if it redistributes income in favour of those with a higher propensity to consume domestically produced goods (as opposed to imports), thus raising aggregate domestic demand. But again the existence of landless labourers complicates the matter beyond a simple comparison of rural and urban or agriculture and industry, because the rural landless most likely have an even greater marginal propensity to consume out of domestic production than do farmers, who in turn are less likely to consume imports than urban dwellers. There is a large literature that argues that agricultural growth is better for economic growth than growth in other sectors, but for this literature to be relevant to the present discussion it would be important to distinguish between agricultural growth that is generated through adoption of new technologies and agricultural growth (if any) that is generated through simple income redistribution. Indeed, in the absence of promising new technologies, there is a potential for price protection to lock farmers into a low-growth crop. Indeed, the Philippines and Sri Lanka, the two developing countries where protection of rice has been the most pronounced for a long period of time, have had the lowest agricultural growth rates among major rice-producing Asian countries since 1980.

A further consideration is that in Asia, agricultural protection and high food prices may reduce the competitiveness of unskilled labour via a wage good argument. This could slow growth and hinder movement of the economy towards sectors in which it has a medium-term comparative advantage, which for many developing countries may be in the realm of labour-intensive manufactures. Of course, there are many factors that affect economic growth, but it is interesting to note that the most dynamic developing countries in Asia seem to have followed a policy of low food prices, while the more stagnant economies have suffered from relatively high food prices.

For example, China, Thailand and Viet Nam have all pursued a policy of keeping rice prices at world market levels for fifteen or twenty years, during which time all experienced quite rapid growth. Indonesia did the same during its period of rapid growth that came to an abrupt end with the onset of the Asian financial crisis. During the past few years, in Indonesia rice prices have been above import parity levels, although clearly this is not the main reason for the slowdown in economic growth. On the other hand, real GDP per capita in the Philippines has been stagnant for the past 20 years, during which time rice prices have been highly protected and well above levels in neighbouring developing countries. Sri Lanka has also had high levels of protection for its rice sector during this time (Weerahewa 2004, p. 51; comparing wholesale prices with export prices from Bangkok). Its growth performance has been substantially better than that of the Philippines, with a doubling of per capita income from 1980 to 2002. This performance was not nearly as strong, however, as that of China, Thailand, Viet Nam and, before the crisis, Indonesia. One might argue that China, Thailand and Viet Nam all went through taxation-removing liberalization for the agricultural sector, and that this liberalization stimulated growth. However, the point here is that rice prices in these countries were always at or below world levels. Thus, one cannot claim that protection of rice production spurred economic growth, because these countries did not protect their rice sectors. In addition, Thailand had already experienced decades of rapid growth before taxation of rice farmers was removed.

Agricultural protection may be justified in some circumstances

While agricultural protectionism has some difficulties associated with it, there may be circumstances in which it is justified. For example, what if the agricultural sector of country X is not competitive at world prices for any crops? If liberalization goes farther, it is possible that some countries might find themselves in this situation, especially in the presence of OECD subsidies that artificially lower world prices (although there is much dispute about the size of this effect). The theory of comparative advantage states that every country must have a comparative advantage in something, given appropriate exchange rates. But that comparative advantage need not be found in the agricultural sector. If in fact country X has no comparative advantage in agriculture, how does one go about moving all the workers out of that sector into other sectors? Agriculture is the single largest employer in most LDCs, so this presents a major problem. Clearly the process will take decades at a minimum, so what are agricultural workers to do in the meantime? If labour markets are imperfect, and farmers and agricultural workers are simply thrown out of work without alternative forms of employment, then even aggregate GDP will not necessarily increase in response to liberalization (never mind the distributional consequences). These implications for rural to urban migration must be taken into account for the developing countries, which have a much larger proportion of their labour force in agriculture than do the developed countries.

Even if one agrees to a certain level of protectionism for the agricultural sector as a whole, it still presumably makes sense to take account of comparative advantage within the agricultural sector. This would entail a positive across the board uniform agricultural tariff in order to preserve some absolute advantage while the population is slowly being transferred out of agriculture[8]. One could also make an argument for preferential protection of labour-intensive crops such as vegetables (not necessarily rice) in order to support the income of rural landless labourers, although it would be important to couple this with increased educational opportunities for such households so that they have a viable long-term strategy for exiting poverty. Unfortunately, agricultural protection in actual practice is often skewed in favour of commodities produced by relatively well-to-do farmers (relative to other farmers, that is, not well to do in some absolute sense). Rice farmers in the Philippines and Indonesia receive preferential protection, as do sugar farmers in many countries.

Research needs

The case studies and discussion presented above, on balance, generally point toward or argue for favourable effects of agricultural trade liberalization in the context of Asia. Perhaps this is due to an inherent bias among economists in favour of reporting positive effects of trade liberalization. Or perhaps it is because trade liberalization is indeed generally a good thing. But even when it is, trade liberalization is no different from any other policy or technology (or change) in that there are winners as well as losers. It would be folly to avoid all change that creates losers, because this is a recipe for stagnation that hurts everyone in the long run. Nevertheless, much of the scepticism surrounding trade liberalization is concerned with the negative effects on certain subsets of the population, and in general the economics profession could do a better job of understanding these effects. The vast majority of papers give only passing attention to the people who are harmed by trade liberalization, often in the form of references to the theoretical possibility of lump sum transfers to compensate the losers. But since such compensating lump sum transfers are rarely done in practice, it is important to push the analysis further in order to better understand the effects of trade liberalization on the losers as well as the winners. Admission that some people are hurt by trade liberalization should not derail the entire process if indeed it is beneficial to society.

The literature that is sceptical of trade liberalization, to its credit, does take seriously the effects on losers. For example, FAO (2000) conducted some studies that were an excellent step in this direction, although more research in this area is needed that takes into account all important effects, not just the negative effects on farmers. For example, it is important to understand the benefits for consumers and the micro-economics of adjustments made by farmers and traders, many of whom may shift their efforts and resources from one crop to another as trade policy changes, thus cushioning the negative impacts.

There is also a lack of rigour in many criticisms of trade liberalization. Indeed, many of these writings contain major factual inaccuracies. Kwa (2000) claims that in the Philippines, due to a focus on export crops, "acreage planted to rice and corn have (sic) been slashed by more than half, from 5 million hectares to 1.9 million hectares". Although no dates are given, data from FAO (2006) show that combined area planted to rice and corn in the Philippines has exceeded 5 million hectares every year since 1961, with the exception of 4.99 million hectares in 1963. In the year 2000, when the statement was written, this combined area exceeded 6.5 million hectares. As another example, the International Forum on Globalization (2003), in complaining about WTO agreements mandating the Philippines to import a small percentage of its rice consumption, claims that the Philippines "has been self-sufficient in rice for centuries", when in fact the Philippines has been consistently importing rice for more than a century (see Rose (1985) and FAO (2006) for supporting data). As one final example, one of the FAO country case studies mentioned above stated that trade liberalization for potatoes and onions in Sri Lanka affected 300,000 persons involved in the production and marketing of these crops (FAO 2000). It is now easy to find citations on the web (e.g. Slatter 2003; Kwa 2000) that state that 300,000 jobs were lost, although the original paper actually did not state that jobs were lost, only that "persons were affected". It is hard to imagine that this is a realistic figure for jobs lost – this would have accounted for nearly 4% of the country's economically active population at the time from a pair of crops that combined account for less than 1% of the total agricultural area in the country. There is also no mention of any empirical assessment of what crops many of these farmers may have planted instead of potatoes and onions (the area planted to potatoes in Sri Lanka increased by a factor of ten in the thirty years prior to liberalization, which at least suggests there are alternative crops). This is not to trivialize the jobs that may indeed have been lost, or the hardship that some farmers suffered, but the analysis must be more rigorous if we are to truly understand the process.

There is a wide range of research that could be conducted to examine the possible negative effects of trade liberalization. For example, ability to adapt to change will be a crucial determinant of just how much damage is suffered by farmers who are hurt by more imports. Certainly the type and quality of land will play an important role, but what about wealth, education and social class? Do wealthier, more educated farmers switch crops more easily in response to changing prices? Just as the Slutsky matrix of demand response to price changes is not the same for all consumers (Timmer 1981), one might expect that the supply response of poor and rich farmers to price changes wrought by trade liberalization will be different. Out of necessity, the poor may be quicker to switch crops when prices change. Or due to lack of human and financial capital, they might not have this flexibility. Most likely, the answer will be location and situation specific. On the consumption side, who gets the benefits of liberalization, the rich or the poor? Does trade liberalization hurt or help the agro-industrialization process, which is a potential source of jobs and learning by doing? Again, the answers will be commodity and country specific.

More research on the stability of international trade is also needed, instead of just assuming that more liberal trade will automatically lead to more stable markets. Returning to the above example from Sri Lanka, India imposed a ban on onion exports in 1998 that led to a quadrupling of prices in Sri Lanka (FAO 2000) because there had been a 25% reduction in the area planted to onions compared with 1997. On the other hand, large price fluctuations for vegetables are not uncommon even without trade liberalization, and the reduced area planted in Sri Lanka in 1998 was still more area than was planted between 1993 and 1995 (i.e., area planted in 1996 and 1997 was unusually high). Thus, the role of India's export ban is not necessarily clear. More rigorous research is needed to understand the many effects of trade liberalization and the research needs to fairly consider both positive and negative outcomes.

NOTES

[1] Some of the work on this paper was carried out while the author was at the International Rice Research Institute (IRRI). The views in this paper represent the views of the author, and not necessarily those of the Food and Agriculture Organization of the United Nations or IRRI. Helpful comments from participants at the Babcock Workshop on 'Agricultural Trade Liberalization and the Least Developed Countries' and from Niek Koning in particular are gratefully acknowledged. The usual disclaimer applies.

[2] Trends in trade are measured for exports and imports separately, each as a share of either production (for net exporters) or as a share of consumption (for net importers).

[3] For estimates of the effects of these reforms in China, see, among others, McMillan et al. (1989), Lin (1992), Kalirajan et al. (1996) and Zhang and Carter (1997). For Viet Nam, see Pingali and Xuan (1992) and Che et al. (2001).

[4] Price changes may possibly be compensated by changes in production, so that income risk is negligible even when prices are fluctuating. If the demand elasticity is greater than 0.5 in absolute value, this compensation will occur at the macro-level in response to supply shocks (although it will not if demand is sufficiently inelastic). However, it is not clear this compensation will occur at the micro-level of individual farmers if supply shocks are asymmetrically distributed across farmers, e.g. some farmers suffer large production shocks while others suffer none. For example, higher prices do not stabilize income if there is no harvest that can benefit from the higher prices. Further, many farmers seem to worry more about market risk, which can be perceived as man-made, than about weather risk, which seems to be more an act of (super)natural origin.

[5] The discussion on fertilizer and irrigation reforms in Bangladesh is based largely on Hossain (1996).

[6] The discussion in this section is based on Gisselquist et al. (2002).

[7] On the other hand, the landless may benefit from higher real wages if the protection causes an increase in the demand for unskilled labor (e.g. Ravallion 1990). However, Rashid (2002) argues this is no longer true for Bangladesh today, although it was in earlier years. Dawe (2003) argues that protection for rice is unlikely to create higher real wages for unskilled farm labourers in the Philippine context.

[8] An undervalued exchange rate would be similar, but not identical, in its effects because most agricultural output is tradable, at least in theory. However, government trade policies often transform tradable goods into non-tradables. Furthermore, there are many tradables outside the agricultural sector that benefit from an undervalued exchange rate.

REFERENCES

Balisacan, A.M., 2000. *Growth, inequality and poverty reduction in the Philippines: a re-examination of evidence*. University of the Philippines, Quezon City.

Bilaterals.org, 2006. *Bilaterals.org: everything that's not happening at the WTO*. Available: [http://www.bilaterals.org/] (February 2006).

Che, T.N., Kompas, T. and Vousden, N., 2001. Incentives and static and dynamic gains from market reform: rice production in Vietnam. *The Australian Journal of Agricultural and Resource Economics,* 45 (4), 547-572.

Dawe, D., 2002. The changing structure of the world rice market, 1950-2000. *Food Policy,* 27 (4), 355-370.

Dawe, D., 2003. Equity effects of rice trade liberalization in the Philippines. *In:* Mew, T.W., Brar, D.S., Peng, S., et al. eds. *Rice science: innovations and impact for livelihood: proceedings of the International Rice Research Conference, 16-19 September 2002, Beijing, China.* International Rice Research Institute, Beijing.

Dawe, D., 2004. Rice to the tiller. *Rice Today,* 3 (3), 41. [http://www.irri.org/publications/today/pdfs/3-3/facts3-3.pdf]

Dawe, D. and Slayton, T., 2004. *Changing structure, conduct, and performance of the world rice market: FAO Rice Conference, Rome, Italy, 12-13 February 2004.* FAO, Rome. [http://www.fao.org/rice2004/en/pdf/dawe.pdf]

Del Ninno, C. and Dorosh, P.A., 2001. Averting a food crisis: private imports and public targeted distribution in Bangladesh after the 1998 flood. *Agricultural Economics,* 25 (2/3), 337-346.

Dorosh, P., 2003. Agricultural reform in Asia. *In: Trade reforms and food security: conceptualizing the linkages.* Food and Agriculture Organization of the United Nations, Rome, Chapter 13. [http://www.fao.org/DOCREP/005/Y4671E/y4671e0j.htm]

Dorosh, P.A., 2001. Trade liberalization and national food security: rice trade between Bangladesh and India. *World Development,* 29 (4), 673-689.

Dowlah, C.A.F., 2003. Bangladesh. *In:* Ingco, M.D. ed. *Agriculture, trade, and the WTO in South Asia.* World Bank, Washington.

FAO, 2000. *Agriculture, trade and food security issues and options in the WTO negotiations from the perspective of developing countries. Volume II: Country case studies.* FAO, Rome. [http://www.fao.org/docrep/003/x8731e/x8731e00.htm]

FAO, 2005. *The state of food and agriculture 2005.* FAO, Rome. FAO Agriculture Series no. 36. [ftp://ftp.fao.org/docrep/fao/008/a0050e/a0050e_full.pdf]

FAO, 2006. *FAOStat: FAO statistical databases.* Available: [http://faostat.fao.org/] (2006).

Gisselquist, D., 2002. Deregulating the transfer of agricultural technology: lessons from Bangladesh, India, Turkey, and Zimbabwe. *The World Bank Research Observer,* 17 (2), 237-265.

Hossain, M., 1996. Agricultural policies in Bangladesh: evolution and impact on crop production. *In:* Abdullah, A. and Khan, A.R. eds. *State market and development: essays in honor of Rehman Sobhan.* University Press, Daka.

Ingco, M., 1995. *Agricultural trade liberalization in the Uruguay Round: one step forward, one step back? Supplementary paper for the Conference on Uruguay Round and Developing Countries, Washington DC, January 26-27, 1995.* World Bank, Washington.

International Forum on Globalization and Center for Food Safety, 2003. *WTO food and agriculture rules: briefing on critical issues.* International Forum on Globalization, San Francisco. [http://www.ifg.org/pdf/int'l_trade-W=_food_&_ag_ru.pdf]

Kalirajan, K.P., Obwona, M.B. and Zhao, S., 1996. A decomposition of total factor productivity growth: the case of Chinese agricultural growth before and after reforms. *American Journal of Agricultural Economics,* 78 (2), 331-338.

Kwa, A., 2000. *The agreement on agriculture: change requires a hero's journey: paper presented Seminar on the Agreement on Agriculture co-organised by the South Centre, Institute for Agriculture and Trade Policy (IATP), Action Aid and Focus on the Global South.* [http://www.focusweb.org/publications/2000/The%20Agreement%20on%20Agriculture.htm]

Landes, R. and Gulati, A., 2004. Farm sector performance and reform agenda. *Economic and Political Weekly,* 39 (32), 3611-3619.

Lin, J.Y., 1992. Rural reforms and agricultural growth in China. *The American Economic Review,* 82 (1), 34-51.

McMillan, J., Whalley, J. and Zhu, L., 1989. The impact of China's economic reforms on agricultural productivity growth. *The Journal of Political Economy,* 97 (4), 781-807.

Mellor, J.W., 1978. Food price policy and income distribution in low-income countries. *Economic Development and Cultural Change,* 27 (1), 1-26.

Pingali, P.L. and Xuan, V.T., 1992. Vietnam: decollectivization and rice productivity growth. *Economic Development and Cultural Change,* 40 (4), 697-718.

Rashid, S., 2002. *Dynamics of agricultural wage and rice price in Bangladesh: a reexamination.* International Food Policy Research Institute, Washington. Markets and Structural Studies Division Discussion Paper no. 44.

Rashid, S., Cummings Jr., R. and Gulati, A., 2005. *Grain marketing parastatals: why do they have to change now?* International Food Policy Research Institute, Washington. MTID Discussion Paper no. 80. [http://www.ifpri.org/divs/mtid/dp/papers/mtidp80.pdf]

Ravallion, M., 1990. Rural welfare effects of food price changes under induced wage responses: theory and evidence for Bangladesh. *Oxford Economic Papers,* 42 (3), 574-585.

Rose, B., 1985. *Appendix to The rice economy of Asia.* Resources for the Future, Washington.

Sahn, D.E., 1988. The effect of price and income changes on food-energy intake in Sri Lanka. *Economic Development and Cultural Change,* 36 (2), 315-340.

Slatter, C., 2003. *Will trade liberalization lead to the eradication or exacerbation of poverty?* [http://www.cid.org.nz/advocacy/Claire_Slatter_-_CID_Trade_Forum.pdf]

Timmer, C.P., 1981. Is there "Curvature" in the Slutsky Matrix? *The Review of Economics and Statistics,* 63 (3), 395-402.

Timmer, C.P., 1993. Rural bias in the East and South-east Asian rice economy: Indonesia in comparative perspective. *Journal of Development Studies,* 29 (4), 149-176.

Timmer, C.P., 1996. Does Bulog stabilise rice prices in Indonesia? Should it try? *Bulletin of Indonesian Economic Studies,* 32 (2), 45-74.

Trade liberalization, agriculture and small households in the Philippines: proactive responses to the threats and opportunities of globalization, 2004. ANGOC, Quezon City.

Weerahewa, J., 2004. *Impacts of trade liberalization and market reforms on the paddy/rice sector in Sri Lanka.* International Food Policy Research Institute, Washington. MTID Discussion Paper no. 70. [http://www.ifpri.org/divs/mtid/dp/papers/mtidp70.pdf]

Zhang, B. and Carter, C.A., 1997. Reforms, the weather, and productivity growth in China's grain sector. *American Journal of Agricultural Economics,* 79 (4), 1266-1277.

CHAPTER 11

WHAT CAN BE LEARNED FROM THE HISTORY OF DEVELOPED COUNTRIES?

NIEK KONING

Senior lecturer, Agricultural Economics and Rural Policy Group, Wageningen University, Wageningen, The Netherlands

INTRODUCTION

Neoclassical economic theory on international trade holds that liberal trade policies maximize economic welfare. Mainstream development economists add that this is also true in a dynamic sense: such policies would help poor countries to acquire the skills and technology that they need to catch up with rich ones (World Bank 1993). Extending this to farm policy, many economists see agricultural trade liberalization as a pre-condition for pro-poor growth in least developed countries (Aksoy and Beghin 2004; Anderson and Martin 2005; Hertel and Winters 2005; Nash and Mitchell 2005).

This position is underscored by model studies that couple strong convictions with methodological weaknesses. For example, Anderson and Martin (2005) envisage large effects from poor countries reducing their agricultural tariffs. However, whether these are the 'welfare gains' they claim cannot be decided since the distribution among households is unknown[1]. Moreover, their comparative-static model cannot assess the impact on development. This latter is also true for Hertel and Winters (2005), even though these authors include the distribution issue. The few dynamic models that are being made tend to stress endogenous growth effects but ignore poverty traps that can make poor economies dual equilibrium systems. Furthermore, there are hardly any studies that point to the impact that tariff reduction in developed countries would have on the least developed countries specifically – a remarkable fact, for even standard models show that these countries

N. Koning and P. Pinstrup-Andersen (eds.), Agricultural Trade Liberalization and the Least Developed Countries, 197–215.

would lose rather than gain since their preferential access to developed country markets would be eroded (Panagariya 2005; Yu, this volume).

Meanwhile, economists who believe that agricultural trade liberalization would generally benefit least developed countries are faced with some realities that seem to belie this notion:

- Many developed countries did not liberalize their agricultural trade during the early stages of their industrialization but protected their farmers, and newcomers like Korea and Taiwan have followed their example. Neoclassical economists assert that agricultural protection harmed poor consumers and retarded growth (E.G. Diao et al. 2002b; Tracy 1989), but I will argue that this is not always clear.
- Most Asian developing countries with successful green revolutions stabilized or supported their agricultural prices at the time these revolutions occurred (Dorward et al. 2002). These cases include countries with rapid growth like Indonesia and Malaysia (Dawe 2001; Jenkins and Lai 1991; Timmer 2002). In Vietnam and Chile, where rapid growth was coupled with the liberalization of agricultural trade, this involved the removal of negative protection rather than reduction in positive protection (Benjamin and Brandt 2002; Valdés et al. 1991)[2].
- Most least developed countries that are caught in stagnation have not protected their agriculture. Development economists blame their situation on 'urban bias' leading to over-taxation of farmers (Bates 1981; Ng and Yeats 1998; World Bank 1981). Yet a country like Kenya, which was praised for being relatively free from these bogeys (Bates 1989), also slipped into the morass, raising doubts about whether domestic factors offer a full explanation.

These anomalies do not refute the urgent need for reforming the multilateral system of agricultural trade, nor do they mean that all liberal reform is bad. However, they do suggest that the real world is more complex than the standard economic model. Rather than bombarding the public with model studies in a bid to confirm preconceived ideas, economists would do better to pay more attention to the empirical lessons told by actual history – the real laboratory of the social sciences. As a first step in this direction, in the next session, I will survey the historical experiences of developed countries with agricultural free trade or protection. I do not present a sophisticated quantitative analysis, but simply point out some major facts and conjunctures. Even if this does not allow me to make absolute statements on causality, it reveals a number of cases that cannot readily be explained by the standard model. In Section "*Experiences with agricultural free trade and protection*", therefore, I reconsider the issue of market failure in agriculture. In Section "*What does it mean for poor countries?*", I discuss policy implications for the least developed countries, focusing particularly on the situation in many countries in Sub-Saharan Africa.

EXPERIENCES WITH AGRICULTURAL FREE TRADE AND PROTECTION

Evolution of agricultural trade policies since the industrial revolution

The Industrial Revolution that started in Britain around 1800 and spread to Belgium and France after the Napoleonic Wars was followed by a period of international liberalization of agricultural trade. The protectionist Corn Laws in Britain were moderated in the 1830s and phased out in 1846-49. This was followed by a liberalization of agricultural trade policies in other countries, especially after the British-French trade treaty in 1860[3]. The subsequent events seem to support the accepted theory. In Britain, large farms bought new fertilizers, feeds, drainpipes and machines to innovate and intensify their production. In other places, British demand for food and farm-based materials stimulated the growth of farm export sectors. In the Southern US, cotton plantations flourished (Fogel and Engerman 1974), and the same was true of large grain farms in East Elbian Germany (Koning 1994 and literature referred to). More generally, agricultural growth interacted with industrial growth. The 'high farming' movement in Britain was coupled with new growth in railways and heavy industry. In Belgium and France, chain and demand linkages of agricultural growth stimulated the continuing of industrialization and – in the US and Germany – its dynamic take-off.

As Figures 1 a-b illustrate for England and the United States, during this episode buoyant demand led to high prices for agricultural products, while farm wages were still largely determined endogenously in rural labour markets. From the late 19th century, however, these conditions changed radically. On the one hand, railways and motor vessels brought new waves of reclamation, while the chemical industry produced cheap fertilizers that accelerated the increase in yields. On the other hand, electricity, internal combustion and artificial fibres led to minerals replacing farm-produced materials on a massive scale. Whereas the latter forces curbed the increase in the global demand for farm products, the former forces boosted the growth in supply, which led to recurrent falls in international agricultural prices (cf. Schultz 1945). Meanwhile, industrial concentration and serial production techniques that allowed a de-skilling of labour increased the industrial competition in labour markets. As a consequence, price declines in agriculture were no longer cushioned by adjustments in farm wages.

The resulting squeeze on farm profits provoked calls for government support from large and small farmers alike. They were backed by manufacturers who feared that rural stagnation would threaten their markets. Under this pressure, liberal farm policies gave way to government intervention, including protection (Koning 1994). According to the standard view, this response would have hampered pro-poor growth and solely been caused by political factors. In this interpretation, the problems of European farmers were caused by a shift in comparative advantage in grains to new countries. In a free market, European agriculture would have adjusted by shifting to livestock or releasing labour to industry (Tracy 1989).

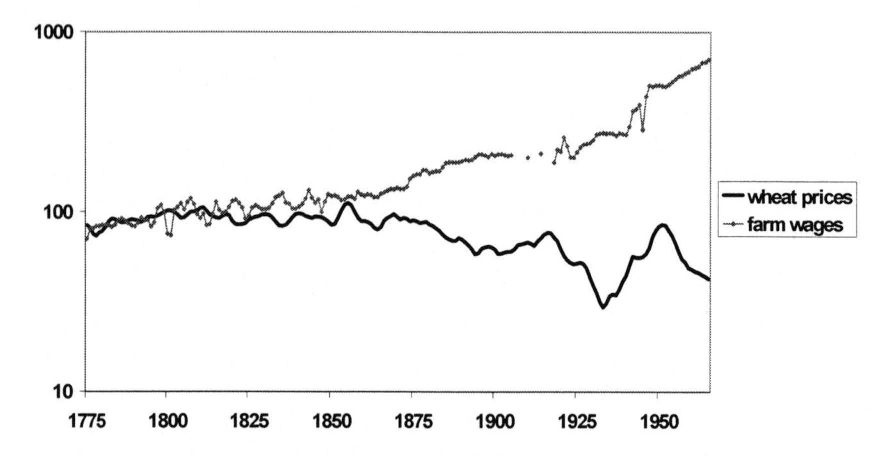

Source: Own calculations based on data in Mitchell (1975, p. 191-5; 736; 1990, p. 737-41; 756-7)

***Figure 1a**. Real wheat prices (5-year moving average) and farm wages, England and Wales, 1818=100*

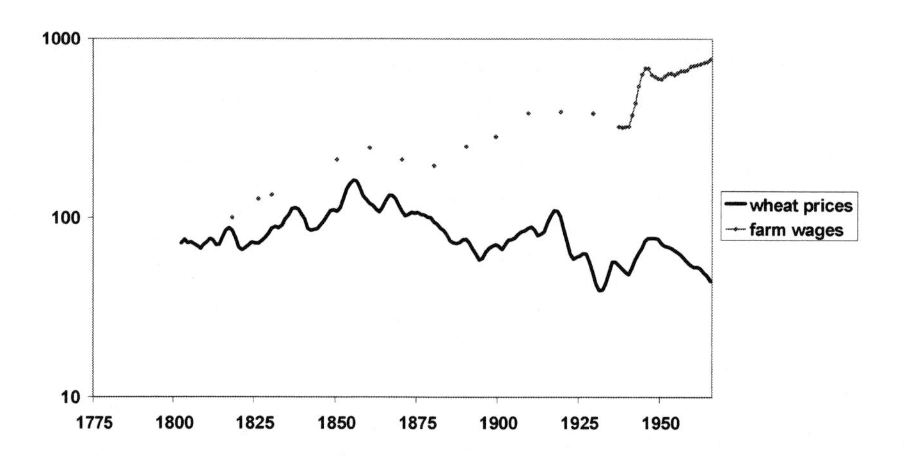

Source: Own calculations based on data in Mitchell (1993, pp. 129-30; 696-8) and US Bureau of the Census (1976, pp. 207-9)

***Figure 1b**. Real wheat prices (5-year moving average) and farm wages, United States, 1818=100*

Could agriculture adjust in a free market?

Cases of successful adjustment
One way to test the above view is to consider the experiences of countries that resisted protection. I will start with the cases that might be seen as supporting the accepted view.

- Most countries in Western Europe protected their farmers from the first fall in agricultural prices, in the late 19th century. However, Denmark, The Netherlands, and the white settler countries across the ocean did not. They weathered the 'agricultural crisis', and when international prices recovered after 1900, dynamic agricultural development resumed (Koning 1994). When prices collapsed again from the late 1920s, however, these countries did resort to protection. Two countries – the US and Denmark – tried to restore free market policies in the 1950s, but they returned to protection after a few years as a price decline later in the decade caused a significant fall in farm incomes. In Denmark, productivity growth was affected, and model studies suggest that the same would have happened in US agriculture had the policy been continued (Cochrane and Ryan 1976; Koning 1986).
- In South Korea and Taiwan, in the 1950s, production and productivity in agriculture increased while output prices were kept below world market levels rather than being supported. The price decline in the later 1950s entailed a slowdown, but in Taiwan agricultural growth resumed after 1960 without protection. South Korea introduced more supportive policies, however, and from the early 1970s, both countries had positive and increasing agricultural protection (Ban et al. 1980; Francks et al. 1999; Moon and Kang 1991).
- After 1984, New Zealand abandoned protection. Although the number of sheep strongly decreased and much marginal hill land went out of production, dairy and horticulture expanded. The adjustment was hailed as a success, not least because it was followed by an increase in productivity growth (Federated Farmers of New Zealand 2002; Johnston and Frengley 1994; Kalaitzandonakes 1994; Sandrey and Reynolds 1990; Sandrey and Scobie 1994). However, this increase was limited to horticulture and may have been due to pre-liberalization investments (Philpott 1994). In the livestock sector, productivity growth remained unaltered in spite of the massive release of marginal resources (ibid.; Lawrence and Diewert 1999; also cf. Cloke 1996; Gibson et al. 1992).

In all the above cases, one finds special advantages in the farm sector:

- The white settler countries around 1900 benefited from abundant fertile land that could be used for extensive export production thanks to new harvesting machines and the Transport Revolution (Koning 1994). Within this group, New Zealand retains especially favourable conditions for dairy and horticulture, with production costs in dairy farming of only half those in prominent dairy countries like the US, Denmark and The Netherlands (IFCN 2003).
- Around 1900, Denmark and The Netherlands were using intensive systems that were on the productivity frontier of European agriculture, while industrial

retardation moderated farm wages. In addition, their livestock systems were well developed and favourably located to supply the growing consumption centres in Britain and West Germany with animal and horticultural products, which were more price-elastic than products like grain (Koning 1994). It has been suggested that the rest of Europe could have followed this example (Tracy 1989). In reality, the international markets for livestock products were soon overstocked by the Dutch and the Danish and some favoured areas outside Europe. A few years after the fall in grain prices in the late 19th century, livestock prices also declined (Bairoch 1976).

- After WWII, South Korea and Taiwan too had low wages and a productive type of intensive agriculture. Before the war, agricultural development had benefited from large public investment in irrigation, research and infrastructure, and from protection at the outer borders of the Japanese Empire. After the war, farmers benefited from the large-scale redistribution of wealth resulting from land reforms, and from massive US aid (Francks et al. 1999).

The cases of the white settler countries around 1900 and New Zealand today are a reminder of the current situations of Cairns Group developing countries like Brazil. The cases of Denmark and The Netherlands around 1900, or Taiwan and South Korea after WWII, are a reminder of some favoured areas in developing countries that are close to urban or export markets and that have become pockets of agricultural intensification (the success story on Machakos District near Nairobi of Tiffen et al. 1994comes to mind). The bottom line is that they were distinctly intra-marginal producers in the global farm economy. To throw some light on the wider evolutionary processes in today's least developed countries, however, it might be more relevant to consider a case where agriculture was closer to the margin but where the government nevertheless kept to free trade. Such a case is Britain between 1880 and 1930.

Agricultural free trade and stagnation in Britain between 1880 and 1930

When international agricultural prices started to fall around 1880, Britain possessed the most technically advanced agriculture in the world. However, strong industrial competition for labour had raised farm wages, and Britain no longer had a comparative advantage in farming. In spite of this, until 1930, a protectionist response was blocked by commercial interests that wanted to maintain the liberal international system and by trade unions that wanted cheap food. According to standard economic theory, free market adjustment might have involved a strong reduction or even total elimination of agriculture. However, if a farm sector managed to survive to some extent, it would see a recovery of profits and productivity growth. In reality, farm profits remained low and productivity stagnated throughout this period. This was not due to a technological ceiling, but to widespread neglect and a drop in investment in new capital goods (see Koning 1994 and literature referred to). Efforts to maintain soil fertility decreased. Two million acres of arable land were turned into grass, but much of it was badly managed. The maintenance of buildings and equipment was neglected, and drainage activity came to a halt. Although farmers bought self-binders to cut down on labour requirements, the demand for more heavy machinery plummeted and Britain's leading position in steam plough construction was lost to Germany and the United States. Several studies agree that throughout this half century, productivity in agriculture stagnated (Koning 1994; O'Gráda 1981; Wade 1981; Van Zanden 1991). Figure 2 shows that by the eve of WWI, British agriculture had fallen far behind the European productivity frontier it had been part of, together with Denmark, The Netherlands and Belgium.

Did agricultural protection hamper pro-poor growth?

Besides asserting that free-market adjustment of agriculture is possible in spite of low world-market prices, the standard view claims that agricultural protection would hamper pro-poor growth. This makes it interesting to consider the cases where rapid economic growth coincided with high agricultural protection. The most important of these are Germany between 1880 and WWI, and South Korea and Taiwan from the 1960s.

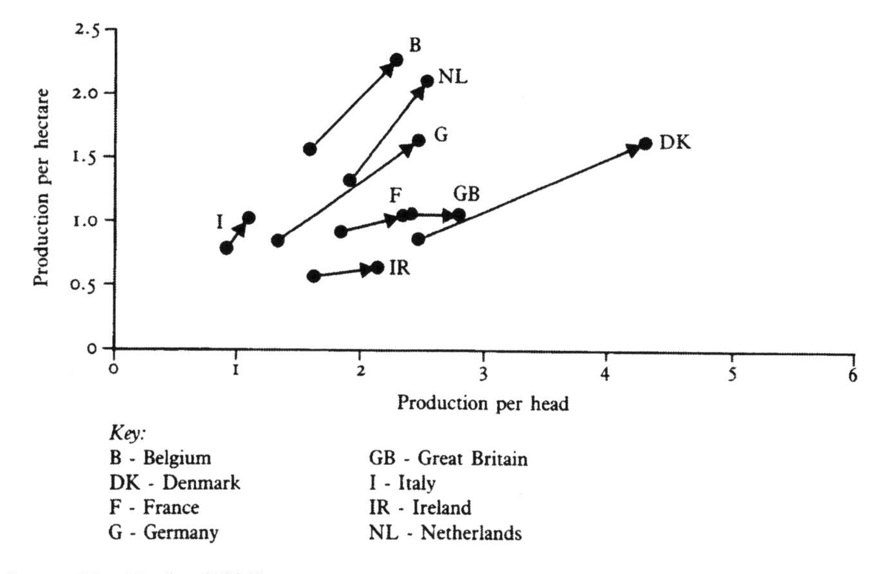

Source: Van Zanden (1991).

Figure 2. *The growth of agricultural productivity per head and per hectare in eight countries of Western Europe, 1870-1910 (in wheat units and 1870 prices)*

Germany between 1880 and WWI. While Britain continued with agricultural free trade until 1930, Germany was the textbook case of protection. When international agricultural prices declined in the 1880s, it raised its farm tariffs sharply. Agricultural protection was moderated in the 1890s by liberal trade treaties, but restored at a high level after the turn of the century.

This policy has been blamed for many problems. While benefiting large grain producers, it would have hampered farm progress and economic growth while raising food prices for poor consumers and feed prices for small livestock farmers (e.g. Gerschenkron 1966; Kempter 1985; Puhle 1986; Rosenberg 1976; Schneider 1987; Tracy 1989). However, economic historians have since revised this view. Although in a static analysis, agricultural protection caused deadweight losses and reduced the buying power of non-farm groups, a dynamic approach shows up matters in a different light. The growth rate of productivity in agriculture was high for European standards (Helling 1966; Koning 1994; Perkins 1981; Van Zanden 1991). From its position far behind the European agricultural productivity frontier, Germany moved to a position close to it (see Figure 2). Even though the poor performance of protectionist France and Italy indicates that protection alone did not guarantee progress, the contrast between the rapid increase in farm productivity in protectionist Germany and its stagnation in free-trading Britain is remarkable. It suggests that more favourable prices may have facilitated investment and innovation in the former country. Of course, the benefits of higher output prices will at least partly have been capitalized in land prices. However, this did not remove the wealth effect for the majority of farmers who already owned their land or had inherited it on conditions that favoured successors. Given the limited mobility of many farmers, the relation of output prices to variable costs was more important than the relation of these prices to total costs. Besides, higher land prices strengthened the ability of these farmers to secure loans.

The idea that agricultural protection would have retarded overall economic growth is unconvincing. Germany was the fastest grower in Europe, and not because it was an advanced industrial country that could 'bear the burden'. When the policy was introduced in the 1880s, the heavy chemical and electrical industries were still in their infancy. Both Webb (1978) and Bairoch (1976) conclude that farm protection accelerated overall growth, allocational distortions being offset by a moderation of emigration and an increase in effective demand[4].

The effect of agricultural protection on the poor has provoked considerable debate. Older assessments have strongly overrated the effect on the costs of living of working class households (Hentschel 1978). Moreover, because agriculture was relatively labour-intensive, agricultural protection will have increased the total demand for labour. This may well have pushed up the real wages of the working poor (Stolper-Samuelson theorem). The contention that agricultural protection hurt small livestock producers has also been refuted (Henning 1987; Moeller 1981; Webb 1982). Livestock production was likewise protected, partly by import restrictions that were justified as sanitary measures. Grain tariffs did not drive up feed costs, because livestock was fed with fodder produced on the farm and with feedstuffs such as oil cakes, which were imported duty-free. Besides, most small farmers were net sellers of grain, and therefore also benefited from grain tariffs.

South Korea and Taiwan after the 1960s

The discussion on agricultural protection in South Korea and Taiwan sounds like a repeat of that on Germany before WWI. Again, the policy is blamed for complicating adjustment of farming structures, retarding economic growth, and harming poor consumers (e.g. Anderson et al. 1986; Beghin et al. 2003; Diao et al. 2002a; 2002b; Van Wijnbergen 1987; Vincent 1989). These contentions are mainly based on partial or general equilibrium model studies. However, the 'welfare losses' indicated by such studies do not allow any strong statements to be made on how poverty or GDP would have evolved over time had farmers not been protected. Indeed, it is quite conceivable that agricultural protection stimulated rather than hampered pro-poor growth. In pre-war Japan, rising rural incomes had been a crucial source of an industrialization that started with the production of simple goods for domestic consumption (Rosovsky and Ohkawa 1961; Ohkawa and Shinohara 1979), and this pattern was repeated in South Korea and Taiwan after WWII (Francks et al. 1999; Park and Johnston 1995). Until the mid-1950s, farmers benefited from American aid, large-scale redistribution of wealth by land reform, and high post-war world market prices that were continued through the Korean War. It allowed increases in farm production and farm incomes that became the driving force behind the development of import substitution industries in this phase. After the Korean War, agricultural prices declined and agricultural development stagnated, entailing a slowdown of industrial growth. This was one of the factors that prompted the governments of both countries to stimulate industrial exports, but the recovery of industrial growth in the 1960s owed as much to a new rise in rural incomes. In Taiwan, this was mainly due to the efficient investment of large amounts of American aid into agricultural infrastructure, research and extension under the aegis of the Sino-American Joint Commission on Rural Reconstruction (Thorbecke 1979). In South Korea, it was also due to the introduction of fertilizer subsidies, import protection, and a strong increase in rice prices paid by the government. This caused a significant improvement of the terms of trade for farmers, even if in real terms protection remained negative because of the overvaluation of the currency (Moon and Kang 1991). When, around 1970, agricultural growth slowed down again, both countries provided positive and increasing protection to their farmers. This was followed by new increases in farm output and incomes, and may well have caused the continuation of agriculture's contribution to the domestic demand pull for industrial growth, even if the relative importance of this contribution declined (also cf. Timmer 1995). As in Germany, it is unlikely that rapid growth was possible only because an advanced industrialization made it possible to bear the burden: especially in South Korea, agricultural protection started when heavy industry was still in its infancy (Francks et al. 1999). Finally, it would be exaggerating to state that protection has frozen agricultural structures. Especially Taiwan is a paragon of successful agricultural diversification. It is true that protection started in rice and, in South Korea, was coupled to government efforts to introduce a new high-yielding rice variety. However, the policy was soon extended to other farm products. Only feeds were less protected, to moderate the costs for domestic livestock producers.

The discussion on agricultural protection is related to that on industrial trade policies. Whereas advocates of 'industrial policy' point to the importance of infant industry protection in the successful industrialization of the two countries (Amsden 1989; Rodrik 1994; Wade 1990), proponents of open-market regimes emphasize the encouragement of industrial exports that facilitated the acquisition of modern technology and skills (Berg and Krueger 2003; World Bank 1993)[5]. However, the precise relations between trade policies, exports and growth are far from clear (Edwards 1993). This leaves room for the hypothesis that agricultural protection, infant-industry protection and the encouragement of industrial exports have reinforced each other rather than conflicting with each other. Agricultural protection maintained the farm contribution to the domestic demand pull for industrialization. Domestic protection of industry prevented this effect from leaking away to other countries through increases in manufactured imports. Both together stimulated industrial growth, which facilitated the cross-subsidization of industrial exports as long as this was still needed to conquer the international markets. In this way, agricultural (and industrial) protection may well have contributed to the advantages of industrial exports that proponents of pro-market policies have emphasized.

POLICY FAILURE OR MARKET FAILURE?

For neoclassical economists, agricultural protection is a mere policy failure. They are convinced that agricultural protection *per se* causes a welfare loss. In their view, economic growth is only related to agricultural protection because higher per capita incomes allow bearing the burden (Anderson et al. 1986). According to these economists, the ubiquity of agricultural protection in developed countries would only be based on the superior political power of agrarian pressure groups – first landed elites, then agribusiness lobbies (Tracy 1989). Referring to Olson's logic of collective action, they explain that the decrease in the numbers of farmers has paradoxically strengthened their ability to organize themselves so as to enforce their interests (Anderson et al. 1986; Schmitt 1984; Senior Nello 1984). Consumers and tax payers, conversely, would be too numerous and heterogeneous to organize countervailing power.

This reasoning is not quite plausible. Two million farmers in the US and seven million in the EU remain too high a number to remove the free rider problem[6]. Moreover, other citizens are not entirely helpless in the face of the agricultural lobbies. In their capacity as workers, employers or voters, they forcefully promote their own interests, setting limits to the extent and government cost of agricultural protection. Besides, as Timmer (1995) has also remarked, nothing in the studies that postulate a causal direction from economic growth to agricultural protection (including diachronic cross-country regressions like Honma and Hayami 1986) contradicts a reverse causation.

Neoclassical economists stick to their explanation because they deny any special problem of low incomes in agriculture. They are convinced that the invisible hand of the market works, by and large, towards an equalization of the earnings of different sectors (Gardner 1992). Conversely, their classical predecessors highlighted the divergence between farm and non-farm earnings (Ricardo 1817). A scarcity of fertilizer combined with high transport costs that prohibited long-distance shipping of staple foods caused population growth to raise the price of farm products, thereby creating a rent for the owners of intra-marginal lands. From the late 19th century, these constraints were broken, but rather than equalizing farm and non-farm incomes, it entailed a reverse divergence. While wages started to increase, the new abundance caused a decline in agricultural prices. In the neoclassical model, the resulting squeeze on farm profits should have prompted a corrective shake-out of small farms and an outflow of farm workers. In reality, limited economies of scale and inherent labour market imperfections weakened these reactions[7]. This was reinforced by an effect that the changes in prices had on farm structures. Rising wages reinforced the advantage that small farms derived from using family labour, while the profit squeeze hindered investment in large farms, thereby eroding their technical advantage. Throughout the western world, large farms declined and family farms increased, increasing the share of self-employed workers who were prone to psychological adaptation to low incomes (Haagsma and Koning 2005). Rather than leaving a depressed sector, as neoclassical theory would predict, many farmers tightened their belts and increased their labour effort in an attempt to defend their incomes by raising their production. Industrial fertilizer, high-yielding seeds, and increased access to markets boosted the influence that these individual responses had on the global supply.

The upshot was the unremitting expansion of agricultural productive capacity that Ray and Harwood describe elsewhere in this volume. Technical progress, land development and reclamation became a treadmill that generated overproduction (see also Cochrane 1958). A balance between the growth in supply and in demand was only achieved when the treadmill squeezed its own fuel supply by reducing farm profits, and thereby investment. As a consequence, free market adjustment did not lead to the efficient equilibrium of neoclassical theory, but rather to a chronic semi-depression in agriculture. This explains the protracted stagnation of productivity growth in agriculture in Britain. Agricultural protection corrected this market failure, at least at the level of national economies. This was what allowed agriculture to increase its productivity and to play its role as a booster of growth in Germany, Korea and Taiwan.

To be sure, protection was not enough to achieve this result. Without the large-scale land reform and the huge investment in infrastructure and in agricultural education and research in these countries, farmers would not have been able to modernize their production to such an extent. (This is illustrated by the sluggish growth of farm productivity in protectionist Italy and France in Figure 2, whose experience may be repeating itself in cases like that of the Philippines, which Dawe describes elsewhere in this volume.) Moreover, agricultural protection caused new distortions in international agricultural markets, including the dumping of surpluses. The only way to prevent these distortions without sacrificing farm progress itself was adequate management of supply. This was envisaged by many proposals and quite some legislation between the 1930s and the 1980s, including the farm laws of the American New Deal, articles 11 and 16 of the GATT, and the production quotas for milk and other products that still exist in the EU and Canada (see, e.g. Benedict 1953; Cochrane and Ryan 1976; Henningson Jr. 1981). However, national egoism and agribusiness expansionism have thwarted most of these attempts. Rather than agricultural protection *per se*, this failure to couple protection to supply management is the real policy failure.

WHAT DOES IT MEAN FOR POOR COUNTRIES?

Around 1900, Britain could afford to sacrifice its agriculture (Koning 1994). As the first 'workplace of the world', its industry had a strong export position. It was less dependent on the home market than the emergent industries of other countries. Also, the share of agriculture in the economy had decreased far more than elsewhere. It eased the absorption of agricultural workers and moderated the impact that low farm earnings had on domestic demand.

In least developed countries like those in Sub-Saharan Africa today, agriculture is stagnating while such mitigating conditions are absent. The effect on economic development is crippling. Poor farmers make poor markets for industry and services. By the same token, the domestic training school for competitive export industries is underdeveloped (cf. Porter 1990). Agrarian malaise leads to a mass exodus from agriculture, but because robust non-agricultural growth is lacking this leads to a proliferation of marginal activities and a jostling for jobs in the public sector.

Neoclassical development economists believe that agricultural stagnation is caused by domestic over-taxation of farmers (Bates 1981; World Bank 1981) or geographic disadvantage (UN Millennium Project 2005). Many blame high transaction costs, weak institutions and bad governance (e.g. Collier and Gunning 1999; Ng and Yeats 1998). They fail to see that these problems are largely a reinforcing feedback effect of stagnation itself. Lack of gainful employment makes people seek refuge in clientelist networks that fight over the distribution of scarce resources. It creates a political market based on the doling out of public sector jobs, which leads to inefficient government services and the undermining of democracy.

Where then should the deeper causes of the stagnation in these countries be sought? Many least developed economies are dual equilibrium systems. When faced with adverse conditions, they fall into a complex poverty trap. The low chain investment trap, soil degradation trap, and macro-economic trap indicated in the papers by Dorward and Kydd, Savadogo, and Ostensson elsewhere in this volume are elements of this. One might also add a low social capital trap: high individual discount rates make people opt for non-cooperative strategies that promise a higher short-term pay-out but that start a vicious spiral of conflict and distrust that hampers productive cooperation in the future (cf. Ostrom 1998).

Pre-industrial European societies were also dual equilibrium systems. 'Agricultural revolutions' alternated with periods when soil degradation and stagnation led to Malthusian crisis (Abel 1978; Grigg 1980; Slicher van Bath 1963). In the former, population growth induced a moderate rise in agricultural prices that stimulated investment and innovation for sustainable agricultural intensification. The resulting growth fuelled the demand for manufactures and services, and strengthened the fiscal base of the state. This upward movement only halted when the techno-institutional capabilities for further adjustments in agriculture were exhausted, so that continued population growth sent agricultural prices skyrocketing. The result was a squeeze on the demand for non-farm goods, making poor farmers over-exploit their plots in an effort to minimize their dependence on food markets, and pushing society into a downward spiral of soil degradation, food insecurity and social disruption, finally ending in demographic stagnation or collapse.

This pre-industrial dynamic was hinged on the endogenous relation between population and agricultural prices (see Figure 3). However, the agricultural treadmill in the world's main farming areas has broken this nexus. It depresses the prices of agricultural products also in regions it bypasses – the more so when industrial countries allow their farm policies to distort world markets. As a consequence, population growth in today's low-income economies fails to provide the price incentives that drove agricultural revolutions in pre-industrial societies, pushing these economies into the poverty trap even though the technical possibilities for sustainable agricultural intensification are far from depleted. Whereas pre-industrial societies fell into crisis when an agricultural revolution was exhausted, in today's low-income economies a similar revolution is nipped in the bud.

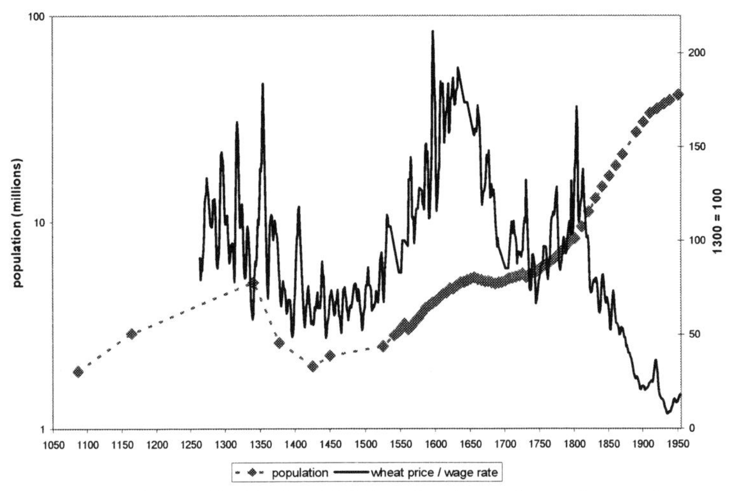

Sources: Population: 1086-1540, Hatcher (1977) and estimates by various authors mentioned in Coleman and Salt (1992, (1992, Table 1.1); 1541-1800, Wrigley et al. (1997); 1801-1954, HMSO (1993). Wheat prices: 1264-1315, Rogers (1866); 1316-1770, Beveridge (1929); 1771-1954, average gazette prices in Mitchell (1990, pp. 756-757). Wages: building wage rates in Phelps Brown and Hopkins (1956).

Figure 3*. Population and ratio between wheat price and wage rate in England, 1086-1954 (population in millions; wheat price / wage ratio as 5-year moving average, 1300 = 100)*

A common objection to this argument is that high internal transport costs limit the influence of international prices on the domestic prices of food crops. However, high transport costs are partly an endogenous factor. Low world market prices have favoured investment in infrastructure that facilitates import rather than internal transport, so that 'price bands' in more remote areas are larger than they otherwise would have been. Indeed, anecdotal evidence suggests that world market prices have had a considerable influence on the evolution of low-income economies. During those decades of the 20th century when international prices were more favourable, agriculture in Sub-Saharan Africa showed considerable dynamism. Conversely, decades with low prices were coupled to stagnation and soil degradation (Koning 2002; Koning and Smaling 2005; Munro 1976). More research on this issue is urgently needed.

How could low-income countries stop themselves from being pushed into a poverty trap by low prices? A first condition is a strong improvement in infrastructure, farm research and the marketing of farm products. Indeed, the increase in public investment needed for this may require generous debt relief and increases in development aid, as Jeffrey Sachs and others are asserting (UN Millennium Project 2005). However, if price ratios remain too unfavourable to allow farmers to invest, more roads and research may bear little fruit and international transfers of means may leak away, as so much development aid has done in the past. In many situations, therefore, supportive price policies may be needed. Many low-income countries in Sub-Saharan Africa have become net importers of food crops, so they could simply protect their farmers through protective tariffs[8]. Ideally, these should be applied at the outer border of regional custom unions with internal free trade to balance national surpluses and shortages, and to allow specialization according to comparative advantage. The tariff revenue could be used for infrastructural projects that could also be used as employment projects to compensate the increased cost of living for poor consumers. Of course, the tariffs should not be too high.

Isn't agricultural protection a regressive taxation of poor consumers? In the first round, yes – like it was in Germany, Japan, Korea and Taiwan, and like the price rises that drove agricultural revolutions in pre-industrial societies also raised the price of bread for the poor. However, if higher prices were to allow investment that raises the demand for labour in and outside agriculture, the net effect for poor consumers may still be positive.

Doesn't agricultural protection raise the price of labour, food being a wage good? To some extent, yes – like it did in the above-mentioned cases, where moderate rises in food prices nevertheless led to effects that fuelled non-farm growth. If protection were to allow agriculture to play its role as an engine of growth, the effects on market demand, skills and social capital may compensate the effect on wages. Empirical research should indicate the size of the tariffs that would achieve an adequate balance between agricultural growth and wage costs.

Doesn't protection provoke the unproductive use of resources for rent seeking? No – because small farmers are too remote from political power to make powerful agricultural lobbies to be feared in poor countries. Doesn't the experience with import substitution industries show that protection breeds inefficiency? No – for the atomistic structure of agriculture ensures more competition between producers. Higher prices do not lead to less but to more efficiency: more innovation, better fertilization of land, and better use of labour – certainly as long as the development of manufacturing remains disappointing (Koning et al. 2001; Reardon et al. 1997; Reardon et al. 1999).

Tariff protection in low-income countries is no panacea. The farm policies of developed countries should also be reformed, as Badiane rightly remarks in this volume. The current substitution of direct payments for price support in the US and the EU is just a shift from one form of nationalistic protectionism to another. In the light of the above analysis, the solution is not multilateral liberalization, but multilateral regulation of the export and import volumes of developed and upper-middle-income countries. However, neither is likely to materialize in the medium term, which makes tariff protection in least developed countries all the more urgent. For this reason, the time has come to lift the taboo on this issue. Empirical research of the effects of price policies should take the place of ideology-based models. Poverty reduction and sustainable growth are too important to be sacrificed to the dogmas of economists.

NOTES

[1] See Jongeneel and Koning (1999), who show that the hypothetical compensation principle that underlies standard welfare economic models cannot reveal actual welfare gains.

[2] Such a move, which also improved price ratios for farmers, heralded the shift to agricultural protectionism in various countries. There are signs that Chile and Vietnam are following suit (see Hachette and Del Pilar Rozas 1993; Nguyen and Grote 2004).

[3] In the United States, farm tariffs were raised again after the Civil War, but the effective level of agricultural protection remained modest.

[4] Bairoch (1976) reached similar conclusions for many other West-European countries. In France, his analysis has been criticized for being global and imprecise. Nevertheless, a critic like Asselain (1985) does not deny that agricultural free trade contributed to the deceleration of growth in that country in the 1870s-80s, and admits that agricultural protection may have been a factor in the recovery in the 1890s.

[5] These issues are less relevant for agriculture, where infant industry problems were less important and exports less important and less knowledge-intensive.

[6] Olson's own remarks on the subject are much more nuanced. He does not speak of farmers having superior power, but of modern communication allowing farmers to catch up finally with urban groups that have long been influential (Olson 1985).

[7] Such imperfections included efficiency wages that introduced entry barriers to industrial labour markets (Akerlof and Yellen 1986) and periodic industrial mass unemployment that locked workers into farming, who were subsequently fixed by the decline in opportunity costs over their lifetime (cf. Johnson and Quance 1972).

[8] A more complex situation would arise if custom unions were to move from being net importers of food crops into net exporters of food crops. Least developed countries have no means to support their farmers through export subsidies or direct payments as developed countries do. The best solution would be a multilateral system of managed trade, in which developed countries (and in a later phase, middle-income countries) would restrict their exports and increase their imports to make room for LDC exports while improving the international prices of agricultural products. Such a system could simply be introduced by imposing maximum quotas on the exports (imports) of developed countries and minimum quotas on imports. These quotas should start from a historical base and could be made tradable between countries to increase flexibility. Of course, although it would technically be quite feasible, the political probability of such a system is low in the medium term. In its absence, least developed countries can only fall back on trade boards with monopoly powers (and on unilateral cartel arrangements for tropical export crops, but this is a difficult road to travel). However, all this is very much for the long term, since most lower-income countries will not become exporters of food crops in the near future.

REFERENCES

Abel, W., 1978. *Agrarkrisen und Agrarkonjunktur: eine Geschichte der Land-und Ernährungswirtschaft Mitteleuropas seit dem hohen Mittelalter*. 3rd edn. Parey, Hamburg.

Akerlof, G.A. and Yellen, J.L., 1986. *Efficiency wage models of the labor market*. Cambridge University Press, Cambridge.

Aksoy, M.A. and Beghin, J.C., 2004. *Global agricultural trade and developing countries*. World Bank, Washington.

Amsden, A.H., 1989. *Asia's next giant: South Korea and late industrialization*. Oxford University Press, New York.

Anderson, K., Hayami, Y., George, A., et al., 1986. *The political economy of agricultural protection: East Asia in international perspective*. Allen & Unwin, Sydney.

Anderson, K. and Martin, W., 2005. Agricultural trade reform and the Doha Development Agenda. *The World Economy,* 28 (9), 1301-1327.

Asselain, J.Ch., 1985. Croissance, dépression et récurrence du protectionnisme francais. *In:* Lassudrie-Duchêne, B. and Reiffers, J.L. eds. *Le protectionnisme: croissance, limites, voies alternatives: colloque du GRECO CNRS EFIQ*. Economica, Paris, 29-53.

Bairoch, P., 1976. *Commerce extérieur et développement économique de l'Europe au XIXe siècle*. Mouton, Paris.

Ban, S.W., Yong, M.P. and Perkins, D.H., 1980. *Rural development: studies in the modernization of the Republic of Korea, 1945–1975*. Harvard University Press, Cambridge.

Bates, R., 1981. *Markets and states in tropical Africa: the political basis of agricultural policies*. University of California Press, Berkeley.

Bates, R.H., 1989. *Beyond the miracle of the market: the political economy of agrarian development in Kenya*. Cambridge University Press, Cambridge.

Beghin, J.C., Bureau, J.C. and Park, S.J., 2003. Food security and agricultural protection in South Korea. *American Journal of Agricultural Economics,* 85 (3), 618-632.

Benedict, M.R., 1953. *Farm policies of the United States, 1790-1950*. Twentieth Century Fund, New York.

Benjamin, D. and Brandt, L., 2002. *Agriculture and income distribution in rural Vietnam under economic reforms: a tale of two regions*. William Davidson Institute, Ann Arbor. William Davidson Working Paper no. 519.

Berg, A. and Krueger, A.O., 2003. *Trade, growth, and poverty: a selective survey*. International Monetary Fund, Washington. IMF Working Paper no. 03/30.

Beveridge, W., 1929. A statistical crime of the seventeenth century. *Journal of Economic & Business History* (1), 503-533.

Cloke, P., 1996. Looking through European eyes? A re-evaluation of agricultural deregulation in New Zealand. *Sociologia Ruralis,* 36 (3), 307-330.

Cochrane, W., 1958. *Farm prices: myth and reality*. University of Minnesota Press, Minneapolis.

Cochrane, W.W. and Ryan, M.E., 1976. *American farm policy, 1948-1973*. University of Minnesota Press, Minneapolis.

Coleman, D. and Salt, J., 1992. *The British population: patterns, trends, and processes*. Oxford University Press, Oxford.

Collier, P. and Gunning, J.W., 1999. Explaining African economic performance. *Journal of Economic Literature,* 37 (1), 64-111.

Dawe, D., 2001. How far down the path to free trade? The importance of rice price stabilization in developing Asia. *Food Policy,* 26 (2), 163-175.

Diao, X., Dyck, J., Skully, D., et al., 2002a. South Korea's agricultural policy hampered economic growth. *Agricultural Outlook*, 10-13.

Diao, X., Dyck, J., Skully, D., et al., 2002b. *Structural change and agricultural protection: costs of Korean agricultural policy, 1975 and 1990*. US Department of Agriculture, Economic Research Service, Washington. AER report no. 809.

Dorward, A., Kydd, J., Morrison, J.A., et al., 2002. *A policy agenda for pro-poor agricultural growth: paper presented at the Agricultural Economics Society Conference, Aberystwyth, 8th-10th April 2001.*

Edwards, S., 1993. Openness, trade liberalization, and growth in developing countries. *Journal of Economic Literature,* 31 (3), 1358-1393.

Federated Farmers of New Zealand, 2002. *Life after subsidies: the New Zealand experience 15 years later*. Federated Farmers of New Zealand, Wellington.

Fogel, R.W. and Engerman, S.L., 1974. *Time on the cross: the economics of American negro slavery*. Little, Brown and Company, London.

Francks, P., Boestel, J. and Kim, C.H., 1999. *Agriculture and economic development in East Asia: from growth to protectionism in Japan, Korea and Taiwan*. Routledge, London.

Gardner, B.L., 1992. Changing economic perspectives on the farm problem. *Journal of Economic Literature*, 30 (1), 62-101.

Gerschenkron, A., 1966. *Bread and democracy in Germany*. Fertig, New York.

Gibson, J., Hillman, J., Josling, T., et al., 1992. *Agricultural and trade deregulation in New Zealand: lessons for Europe and the CAP: paper for the 28th EAAE Seminar, 'EC agricultural policy by the end of the century', Lisboa, September 10-12.*

Grigg, D.B., 1980. *Population growth and agrarian change: an historical perspective*. Cambridge University Press, Cambridge.

Haagsma, R. and Koning, N., 2005. Endogenous norms and preferencesand the farm income problem. *European Review of Agricultural Economics*, 32 (1), 25-49.

Hachette, D. and Del Pilar Rozas, M., 1993. *The liberalization of the Chilean agriculture: 1974-1990*. Catholic University of Chile, Santiago. Working Paper Department of Economics no. 157.

Hatcher, J., 1977. *Plague, population, and the English economy, 1348-1530*. Macmillan, London.

Helling, G., 1966. Die Entwicklung der Produktivität in der deutschen Landwirtschaft im 19. Jahrhundert. *Jahrbuch für Wirtschaftsgeschichte*, 1, 129-141.

Henning, F.W., 1987. Vom Agrarliberalismus zum Agrarprotektionismus. *In:* Pohl, H. ed. *Die Auswirkungen von Zöllen und anderen Handelshemmnissen auf Wirtschaft und Gesellschaft vom Mittelalter bis zur Gegenwart*. Stuttgart, 252-274.

Henningson Jr., B.E., 1981. *United States agricultural trade and development policy during world war II: the role of the Office of Foreign Agricultural Relations*. University of Arkansas, Fayetteville. Ph.D. diss. University of Arkansas

Hentschel, V., 1978. *Wirtschaft und Wirtschaftspolitik im wilhelminischen Deutschland: organisierter Kapitalismus und Interventionsstaat?* Klett-Cotta, Stuttgart.

Hertel, T.W. and Winters, L.A., 2005. *Putting development back into the Doha agenda: poverty impacts of a WTO agreement*. World Bank, Washington.

HMSO, 1993. *1991 Census: historical tables, Great Britain*. HMSO, London.

Honma, M. and Hayami, Y., 1986. Structure of agricultural protection in industrial countries. *Journal of International Economics*, 20 (1), 115-129.

IFCN, 2003. *Dairy report 2002: Status and prospects of typical dairy farms world-wide*. IFCN, Braunschweig.

Jenkins, G. and Lai, A., 1991. Malaysia. *In:* Krueger, A.O., Schiff, M. and Valdés, A. eds. *The political economy of agricultural pricing policy. Vol. 2: Asia*. Johns Hopkins University Press, Baltimore.

Johnson, G.L. and Quance, C.L., 1972. *The overproduction trap in US agriculture: a study of resource allocation from World War I to the late 1960's*. Johns Hopkins University Press, Baltimore.

Johnston, W.E. and Frengley, G.A.G., 1994. Economic adjustments and changes in financial viability of the farming sector: the New Zealand experience. *American Journal of Agricultural Economics*, 76 (5), 1034-1040.

Jongeneel, R. and Koning, N., 1999. The concept of potential Pareto improvement revisited. *Tijdschrift voor Sociaalwetenschappelijk Onderzoek van de Landbouw*, 14 (3), 114-126.

Kalaitzandonakes, N.G., 1994. Price protection and productivity growth. *American Journal of Agricultural Economics*, 76 (4), 722-732.

Kempter, G., 1985. *Agrarprotektionismus: landwirstschaftliche Schutzzollpolitik im Deutschen Reich von 1879 bis 1914*. Lang, Frankfurt am Main. Europäische Hochschulschriften no. 5.

Koning, N., 1986. Beperkingen van landbouwinkomenspolitiek in kapitalistische industrielanden. *Tijdschrift voor Sociaalwetenschappelijk Onderzoek van de Landbouw*, 1, 113-133.

Koning, N., 1994. *The failure of agrarian capitalism: agrarian politics in the UK, Germany, the Netherlands and the USA, 1846-1919*. Routledge, London.

Koning, N., 2002. *Should Africa protect its farmers to revitalise its economy?* International Institute for Environment and Development, London. IIED Gatekeeper Series no. 105.

Koning, N., Heerink, N. and Kauffman, S., 2001. Food insecurity, soil degradation and agricultural markets in West Africa: why current policy approaches fail. *Oxford Development Studies*, 29 (2), 189-207.

Koning, N. and Smaling, E., 2005. Environmental crisis or 'lie of the land? The debate on soil degradation in Africa. *Land Use Policy*, 22 (1), 3-11.

Lawrence, D. and Diewert, E., 1999. *Measuring New Zealand's productivity: report for Department of Labour, Reserve Bank of New Zealand and The Treasury*. New Zealand Treasury. Treasury Working Paper Series no. 99/5-1.

Mitchell, B.R., 1975. *European historical statistics*. Macmillan, Cambridge.
Mitchell, B.R., 1990. *British historical statistics*. Cambridge University Press, Cambridge.
Mitchell, B.R., 1993. *International historical statistics: The Americas 1750-1988*. Stockton Press, New York.
Moeller, R.G., 1981. Peasants and tariffs in the Kaiserreich: how backward were the Bauern. *Agricultural History*, 55 (4), 370-384.
Moon, P.Y. and Kang, B.S., 1991. The Republic of Korea. *In:* Krueger, A.O., Schiff, M. and Valdés, A. eds. *The political economy of agricultural pricing policy. Vol. 2: Asia*. Johns Hopkins University Press, Baltimore.
Munro, J.F., 1976. *Africa and the international economy 1800-1960; an introduction to the modern economic history of Africa south of the Sahara*. Dent, London.
Nash, J. and Mitchell, D., 2005. How freer trade can help feed the poor. *Finance & Development*, 34-37.
Ng, F. and Yeats, A.J., 1998. *Good governance and trade policy: are they the keys to Africa's global integration and growth?* World Bank, Washington.
Nguyen, H. and Grote, U., 2004. *Agricultural policies in Vietnam: producer support estimates 1986-2002*. IFPRI, Washington. Center for Development Research Discussion Paper no. 93.
O'Gráda, C., 1981. Agricultural decline 1860-1914. *In:* Floud, R. and McCloskey, D. eds. *The economic history of Britain since 1700. Part 2: 1860-1939*. Cambridge, 175-197.
Ohkawa, K. and Shinohara, M., 1979. *Patterns of Japanese economic development: a quantitative appraisal*. Yale University Press, New Haven.
Olson, M., 1985. Space, agriculture, and organization. *American Journal of Agricultural Economics*, 67 (5), 928-937.
Ostrom, E., 1998. A behavioral approach to the rational choice theory of collective action. *The American Political Science Review*, 92 (1), 1-22.
Panagariya, A., 2005. Agricultural liberalisation and the least developed countries: six fallacies. *The World Economy*, 29 (8), 1277-1299.
Park, A. and Johnston, B., 1995. Rural development and dynamic externalities in Taiwan's structural transformation. *Economic Development and Cultural Change*, 44 (1), 181-208.
Perkins, J.A., 1981. The agricultural revolution in Germany, 1850-1914. *Journal of European Economic History*, 10 (1), 71-118.
Phelps Brown, E.H. and Hopkins, S.V., 1956. Seven centuries of the prices of consumables, compared with builders' wage-rates. *Economica*, 23 (92), 296-314.
Philpott, B., 1994. *Productivity growth by type of farming: 1972-1993*. Victoria University, Wellington. Research Project on Economic Planning Paper no. 259.
Porter, M.E., 1990. *The competitive advantage of nations*. Macmillan, Basingstoke.
Puhle, H.J., 1986. Lords and peasants in the Kaiserreich. *In:* Moeller, R.G. ed. *Peasants and lords in modern Germany: recent studies in agricultural history*. Allen & Unwin, Boston, 81-109.
Reardon, T., Barrett, C., Kelly, V., et al., 1999. Policy reforms and sustainable agricultural intensification in Africa. *Development Policy Review*, 17 (4), 375-395.
Reardon, T., Kelly, V., Crawford, E., et al., 1997. Promoting sustainable intensification and productivity growth in Sahel agriculture after macroeconomic policy reform. *Food Policy*, 22 (4), 317-327.
Ricardo, D., 1817. *On the principles of political economy and taxation*, London. [http://www.marxists.org/reference/subject/economics/ricardo/tax/index.htm]
Rodrik, D., 1994. *King Kong meets Godzilla: the World Bank and the East Asian miracle*. Centre for Economic Policy Research, London. CEPR Discussion Paper no. 944.
Rogers, J.E.T., 1866. *A history of agriculture and prices in England: from the year after the Oxford parliament (1259) to the commencement of the continental war (1793)*. Clarendon Press, London.
Rosenberg, H., 1976. *Grosse Depression und Bismarckzeit: Wirtschaftsablauf, Gesellschaft und Politik in Mitteleuropa*. Walter de Gruyter, Frankfurt.
Rosovsky, H. and Ohkawa, K., 1961. The indigenous components in the modern Japanese economy. *Economic Development and Cultural Change*, 9 (3), 476-501.
Sandrey, R. and Reynolds, R., 1990. *Farming without subsidies: New Zealand's recent experience*. New Zealand Ministry of Agriculture and Fisheries, Wellington.
Sandrey, R.A. and Scobie, G.M., 1994. Changing international competitiveness and trade: recent experience in New Zealand agriculture. *American Journal of Agricultural Economics*, 76 (5), 1041-1046.

Schmitt, G., 1984. Warum die Agrarpolitik ist, wie sie ist, und nicht, wie sie sein sollte. *Agrarwirtschaft,* 33, 129–136.

Schneider, J., 1987. Die Auswirkungen von Zöllen und Handelsverträgen sowie Handelshemmnissen auf Staat, Wirtschaft und Gesellschaft zwischen 1890 und 1914. *In:* Pohl, H. ed. *Die Auswirkungen von Zöllen und anderen Handelshemmnissen auf Wirtschaft und Gesellschaft vom Mittelalter bis zur Gegenwart.* Stuttgart, 293-327.

Schultz, T.W., 1945. *Agriculture in an unstable society.* McGraw-Hill, New York.

Senior Nello, S., 1984. An application of Public Choice Theory to the question of CAP reform. *European Review of Agricultural Economics,* 11 (3), 261–283.

Slicher van Bath, B.H., 1963. *The agrarian history of Western Europe: ad 500-1850.* Arnold, London.

Thorbecke, E., 1979. Agricultural development. *In:* Galenson, W. ed. *Economic growth and structural change in Taiwan.* London.

Tiffen, M., Mortimore, M. and Gichuki, F., 1994. *More people, less erosion: environmental recovery in Kenya.* Wiley, London.

Timmer, C.P., 1995. Getting agriculture moving: do markets provide the right signals? *Food Policy,* 20 (5), 455-472.

Timmer, C.P., 2002. *Food security and rice price policy in Indonesia: the economics and politics of the food price dilemma.* Bappenas/USAID/DAI Food Policy Advisory Team. Indonesian Food Policy Program Working Paper no. 7.

Tracy, M., 1989. *Government and agriculture in Western Europe, 1880-1988.* New York University Press, New York.

UN Millennium Project, 2005. *Investing in development: a practical plan to achieve the Millennium development goals.* Earthscan, London. [http://www.unmillenniumproject.org/reports/fullreport.htm]

Valdés, A., Hurtado, H. and Muchnik, E., 1991. Chile. *In:* Krueger, A.O., Schiff, M. and Valdés, A. eds. *The political economy of agricultural pricing policy. Vol. 1: Latin America.* Johns Hopkins University Press, Baltimore.

Van Wijnbergen, S., 1987. Short-run macroeconomic effects of agricultural pricing policies. *In:* Newbery, D. and Stern, N. eds. *The theory of taxation for developing countries.* Oxford University Press, Oxford, 521-530.

Van Zanden, J.L., 1991. The first green revolution: the growth of production and productivity in European agriculture, 1870-1914. *The Economic History Review,* 44 (2), 215-239.

Vincent, D., 1989. Domestic effects of agricultural protection in Asian countries with special reference to Korea. *In:* Stoeckel, A.B., Vincent, D. and Cuthbertson, S. eds. *Macroeconomic consequences of farm support policies.* Duke University Press, Durham, 150-172.

Wade, R., 1990. *Governing the market: economic theory and the role of government in East Asian industrialization.* Princeton University Press, Princeton.

Wade, W.W., 1981. *Institutional determinants of technical change and agricultural productivity growth: Denmark, France and Great Britain, 1870-1965.* Arno Press, New York.

Wattenberg, B.J., 1976. *The statistical history of the United States from colonial times to the present.* Basic Books, New York.

Webb, S.B., 1978. *The economic effects of tariff protection in Imperial Germany, 1879 to 1914.* University of Chicago, Chicago. PhD thesis, University of Chicago

Webb, S.B., 1982. Agricultural protection in Wilhelminian Germany: forging an empire with pork and rye. *Journal of Economic History,* 42 (2), 309-326.

World Bank, 1981. *Accelerated development in Sub-Saharan Africa.* World Bank, Washington.

World Bank, 1993. *The East Asian miracle: economic growth and public policy.* Oxford University Press, Oxford.

Wrigley, E.A., Davies, R.S., Oeppen, J.E., et al., 1997. *English population history from family reconstitution 1580-1837.* Cambridge University Press, Cambridge.

CHAPTER 12

HOW U.S. FARM POLICIES IN THE MID-1990S AFFECTED INTERNATIONAL CROP PRICES

A harbinger of what to expect with further world-wide implementation of WTO-compliant policy modifications?

DARYLL E. RAY AND HARWOOD D. SCHAFFER

Agricultural Policy Analysis Center, University of Tennessee, Knoxville, Tennessee, USA

THE NATURE OF THE CROP SECTOR

The belief that the crop agriculture market will self-regulate in the short run is the operative assumption behind the contention that once agricultural subsidies are eliminated, US production will fall and prices will recover. For this assumption to be valid, the market for total crop production must be reasonably responsive to changes in price. That is, the supply and demand price elasticities cannot be exceedingly small in absolute terms.

Considering supply first, farmers tend to plant all of their acres across a wide range of prices. They may change the mix of crops, in an attempt to maximize the revenue per acre, but they almost always plant all of their crop acreage. This farmer behaviour of barely reducing planted acreage in response to dramatically lower prices results in a low price elasticity of supply. Farmers respond to lower prices in this manner, because any dollar earned above the out-of-pocket variable cost of production can be applied to fixed costs like taxes. And on rented ground, the producer has every incentive to use every acre possible. It makes no sense to pay the cost of renting ground, if the intention is to leave it unplanted. The strong tendency of producers to grow crops on every acre is true even as individual farmers go out of business. The land almost universally remains in production just under new management. Crop agriculture tends to use all of its productive capacity all of the time and let the weather determine the final production numbers.

N. Koning and P. Pinstrup-Andersen (eds.), Agricultural Trade Liberalization and the Least Developed Countries, 217–232.

Total crop supply changes little with price; what about the price responsiveness of total food demand? Food is different from other products that consumers purchase. Unlike clothes or televisions, lower prices do not induce consumers to purchase more food. With low television prices, consumers are likely to purchase an extra one for the den and their teenager's bedroom. However, people do not begin to eat four or five meals a day just because food becomes less expensive. In response to lower food prices, consumers may eat out more often, purchase a better quality of food, and buy more highly processed food products, but they do not significantly increase their aggregate food consumption level[1]. The price elasticity of demand for all agricultural products taken together is very low.

If consumers bought more food in response to lower prices and producers cut their total farm output as prices declined, excess inventories would quickly vanish and prices would arrive at profitable levels once again. That is exactly what does not happen. If that self-correction were to occur, there would be no fundamental price and income problem and, therefore, no need for farm programs[2].

THE PRODUCTIVE CAPACITY OF U.S. AGRICULTURE

Aside from random shifts due to weather and other natural events, the downward pressure on crop prices occurs because agriculture's productive capacity tends to expand faster than demand. Demand for agricultural products in a country like the US grows with population and exports but, unlike the demand for cars, houses, clothes and most other product categories, doubling a consumer's income will have a minor impact on his demand for food.

The growth in the productive capacity of US agriculture goes back centuries before the introduction of commodity programs in the 1930s. From its birth as a nation, the US has pursued policies that promoted a phenomenal growth in the productive capacity of agriculture. Supported by the taxpaying public, these developmental policies have increased agriculture's productive capacity by making agricultural inputs more plentiful, more productive or less costly.

Developmental policies began with frontier expansion through the mechanism of land distribution – beginning as early as the late 1700s and continuing through the Homestead Act of 1862. Then, once the frontier closed, the US's most important developmental farm policy was public investment in experiment stations in each state (Hatch Act of 1887), land-grant universities (Morrill Act of 1862), and the cooperative extension service (Smith-Lever Act of 1914). This set of institutions increased the supply, lowered the cost, and improved the quality of physical inputs like seed, chemicals, equipment, and of less tangible inputs like the managerial and decision-making abilities of farmers. The mammoth growth in agricultural productive capacity in the US has been and still is the result of the continuous public investment in agricultural research and education. Clearly, the US government has been intervening in agricultural markets in a gargantuan way for well over a century, expanding productive capacity separate from any consideration of 'farm program' subsidies.

Since individual farmers cannot influence prices, they are constantly in search of new ways to lower costs or increase yields. Thus, the agricultural sector quickly adopts productivity-enhancing technologies, which typically increase supplies faster than the growth in demand, thus putting downward pressure on prices. The lower prices, in turn, become further incentive to adopt more cost-reducing technologies, and prices continue their slide. But the lower prices typically do not cause individual farmers to cut back significantly on production. As a result, production agriculture is under constant price pressure, with periods of brief reprieves, generally the result of disasters or other random events. And, unlike other economic sectors, when individual farmers are forced out of business, the resource (land) is not converted to another more profitable use but remains in production[3].

From a societal perspective, ensuring that all people have access to an abundant, safe, affordable supply of food and fibre is an important public policy objective. Given the typically long research cycles for technology development and the uncompromising daily need for food, public research to improve agricultural productivity and expand agriculture's productive capacity will likely remain an important policy priority in the future.

NEED TO MANAGE PRODUCTIVE CAPACITY

The ability to produce in excess of current needs, however, is not a mandate for agriculture to use all of its productive capacity all of the time. For example, in the manufacturing sector, between 15 and 25% of productive capacity is intentionally idled at any given time by reason of market supply and demand conditions (Economic Report of the President 2003). For the manufacturing sector, actions taken by producers and consumers are generally sufficient to enable self-correction when the sector experiences periods of overstocks and depressed prices. Management of its productive capacity is a critical part of the correction process.

In the face of the relatively low price elasticities of supply and demand one of the challenges for the agriculture sector has been to find ways to manage its productive capacity so as to provide for the food needs of the consuming public while at the same time ensuring a price that covers costs or most costs. Attempts at self-management by farmers, from Henry A. Wallace's newspaper campaign to convince farmers to plant 10% of their acreage to alfalfa (Culver and Hyde 2001, p. 56), to the Farmers Holiday Association (Shover 1965), to the holding actions of the National Farmers Organization (Halcrow et al. 1984, p. 25), have been unsuccessful. Individual farmers are too small for their decisions to have an effect on total production but, unless a large percentage of farmers participate in a voluntary self-management program, the program is likely to fail. Historically, farmers have not been successful in organizing self-help supply management schemes to adjust output to the needs of the market.

As a result, the role of managing the ever-increasing productive capacity of agriculture has fallen to the government. Wallace and other New Dealers designed the original commodity programs to do for agriculture what it could not do for itself but other industries do on a regular basis: manage productive capacity to provide sustainable and stable prices and incomes. Until the mid-1980s – and beyond, in some cases – the primary focus of US commodity policy was on production management programs and price support and stabilization programs.

These programs worked by vesting the Secretary of Agriculture with a variety of tools that could be used to manage the productive capacity of US agriculture in the same way that a corporate CEO manages the productive capacity of her firm. These tools established a set of bounds within which the free market could work to allocate resources. Despite their built-in complications, supply management policies have historically prevented the chronic overproduction and depressed prices that would have occurred from a full use of agriculture's productive capacity all of the time. Price support programs put a floor under major-crop prices. So if the Secretary erred in setting aside too little acreage because of above-average yields or unusually low demand, prices were prevented from plummeting uncontrollably.

THE CHANGE IN U.S. POLICY IS THE REASON FOR LOW PRICES

Over the last two decades, the US policy goal of ensuring growth in productive capacity has remained, but the goal to protect prices and farmer incomes through managing the capacity has not. Rather, the government has placed its reliance on the free market to determine prices, making direct payments to support farmer incomes during times of low prices. To absorb excess inventory, US policy shifted away from production management and price support and toward demand expansion – especially export demand. Advocates of freer markets and trade liberalization were successful in persuading policy makers to encourage lower prices by reducing crop price supports, expecting that a flood of exports would follow. It was predicted that by modifying the 'government intervention' of price supports, other countries would reduce their production, higher prices would return, and farmers would reap the benefits of this export boom.

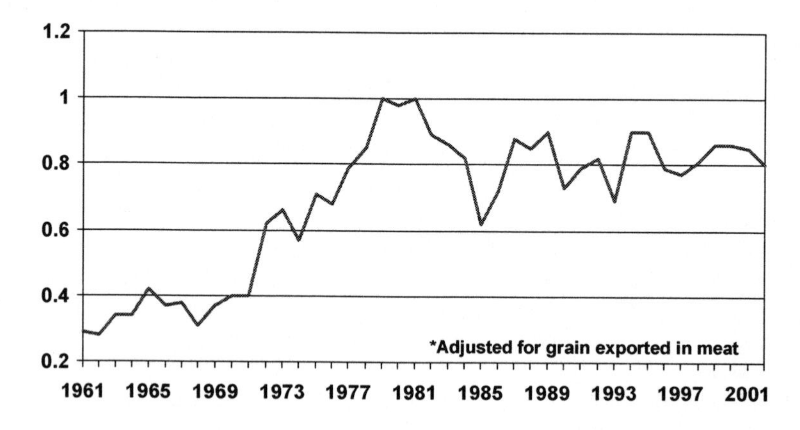

Figure 1. *Index of US 8 crop exports, 1961-2002, adjusted for grain exported in meat, 1979=1.0*

The result of this kind of thinking was the 1996 Farm Bill, which removed all vestiges of government price supports and annual supply controls. At the time, USDA forecasters were projecting a tremendous growth in US crop exports for the foreseeable future. As shown in Figure 1, the aggregate trend of US exports for the eight major crops continued to be flat after 1996. The skyward export trend in the 1970s, while perhaps burned into people's minds, does not reflect the reality of the last quarter century.

The removal of the set-aside program freed up acreage previously withheld from production. Thus, it was no surprise that acreage planted to the eight major crops increased over 6% (over 15 million acres) the year the set-aside policy was removed. Inventory adjustments and world conditions staved off massive price declines, but only until 1998. Thereafter prices plummeted, and government subsidies ballooned to compensate for lost market income. Even as prices declined, the previously idled acreage – which came into production in 1996 – remained in cultivation. While the indexed market price for the eight major crops declined by nearly 40%, contrary to expectations, these radically lower prices did not appreciably cut the aggregate crop acreage remaining in use (Ray et al. 2003, p. 19).

Another feature of the 1996 policy – the effective elimination of price supports – has had the effect of sustaining the persistence of low prices. Current US agricultural policy has no tools to limit the downward price spiral. Even successive yearly reductions in grain stocks have not had the expected price-enhancing impacts of the past. In the current environment, market participants know that no supply management programs can be used next year to raise prices. As a result, crop demanders do not bid up prices to secure future grain needs. They rightly expect, with all-out production, prices will be as low or lower next season. Over the last five years, market participants have been more and more comfortable with less and less grain in the granary at the end of the crop year. Hence, prices have fallen much further than they would have under similar stock conditions before 1996 (Schaffer 2004; unpubl.).

Prior to 1996, government commodity payments were generally used as financial incentives to encourage farmers to participate in supply management programs. Since 1996, government commodity payments are strictly income support payments. In response to the massive price slide, Congress instituted record level payments to farmers, partially compensating farmers for lost income. Annual commodity payments by program are presented in Figure 2. Beginning in 1998, subsidies to farmers increased by 250% over the 1990-1997 period. Post-1997 subsidies took the form of direct payments, unanticipated loan deficiency payments (LDPs), marketing loan gains, and ad hoc/emergency/disaster payments. While all of these payments were being made, the so-called 'excess stocks' were not being removed from the marketplace. As a consequence, prices plummeted. But, with a low price elasticity of demand, falling prices did not cause consumers to buy and consume the problematic 'excess stocks.' As a result prices did not recover but continued their free fall.

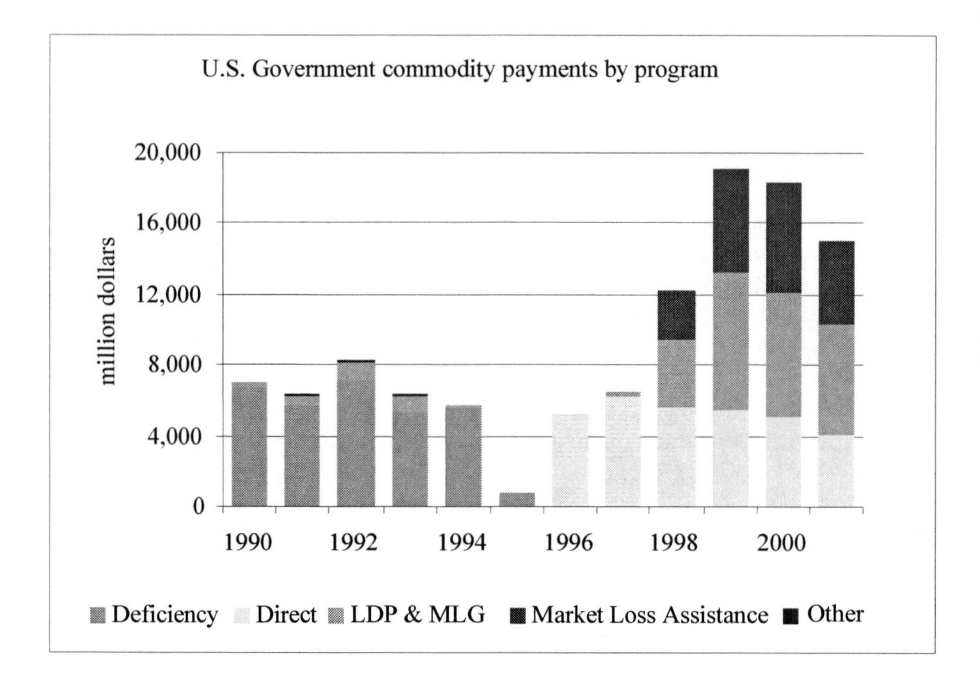

Figure 2. *US Government commodity payments by program, 1990-2001*

The resulting low prices triggered high subsidies in the US, *not* the reverse. While some blame high US subsidies for low prices, the data clearly show the opposite: higher and higher subsidies were authorized in response to lower and lower prices and farm incomes. The problem was not caused by the income support payments that were added, *post hoc*, by recent legislation, but by the supply control and price-supporting mechanisms that were taken away. As prices fell – making US commodities appear to be more competitive in world markets – exports remained flat.

IMPACT OF U.S. POLICY CHANGE ON INTERNATIONAL AGRICULTURE

The removal of the price floor in the US had a major impact on farmers around the world as prices descended to levels that were often below the cost of production. The difference between farmers in the US and elsewhere is twofold. First, the US government could afford to and did grant huge payments to farmers to make up for the low prices while most developing countries could not afford those payments. Second, developing countries were unable to use their traditional method, tariffs, to protect their producers from low world prices. The use of tariffs as a means of import/export control had been significantly weakened in many countries as the result of structural readjustment policies forced on these countries by the World Bank and the IMF. Consequently, farmers in developing countries bore the brunt of the failure of the US to maintain its historic supply management program. It was not the payments that were given to US farmers, but rather the attempt by the US to impose a global trade liberalization scheme on US and world agriculture that brought about this disaster.

In summary, recent periods of depressed prices in world crop markets can, in fact, be traced to policy actions taken in the US. But, the origin and nature of those actions may be different from what most would expect. The US continually invests in the creation of new technologies and other means to expand agricultural production. Overproduction, in the economic sense of that term, occurs when the growth in agriculture's productive capacity exceeds growth in demand *and* there is no mechanism in place to throttle production to needs. The drop in crop prices occurred when the US eliminated the means that had long been used to throttle or balance output to demand needs. Because of the nature of food and aggregate agricultural production, neither consumers nor producers appreciably changed the quantity they consumed and produced, respectively. As a result, market self-correction for total agriculture did not/could not occur in a timely fashion.

The US choose to replace a portion of the nation's farmers' reduced market revenue with payments but that action was a result of the low prices, not the cause. Had the payments not occurred, land prices would have declined substantially, some farmers would have sold out to other better-financed farmers, and the land would have largely remained in agricultural use. Total production would have been affected very little, and, hence, the overall level of farm prices would have recovered only slightly.

Hence, the accusation that US agriculture, in economic terms, overproduced and depressed prices worldwide in recent years has merit. While excess productive capacity is typical in the US, most of price depression should be blamed on the elimination of supply control mechanisms. Any increase in US production that is directly attributable to changes in government payment levels is miniscule compared to the effect of eliminating set-asides and price floors.

An examination of the aggregate output effect of reduced subsidies in other countries

The assertion that reducing and/or eliminating government programs will lead to a reduction in production can be tested by looking at the experience of countries that have made changes in these programs. Over the last few decades, several countries have moved toward policies of reducing government involvement in agricultural markets. Canada, Mexico and Australia have established track records of fewer government controls and freer markets. Changes in commodity production in these countries are the result of a complex array of factors. However, evidence clearly indicates that removal of and reductions in subsidies have not led to significant drops in production. In fact, production increased in several cases. These observations are consistent with studies using the IMPACT and POLYSYS models, which showed that eliminating subsidies will not significantly or quickly reduce production or increase prices.

The Canadian experience. Huge increases in Canadian agricultural subsidies through the 1980s contributed to less than a 3% rise in the number of acres cultivated. Then, fiscal deficits in the 1990s forced a 35% cutback in Canada's support programs over a three-year period. The most notable was the elimination of all subsidies for grain transportation in 1995. This and other significant reductions in government support levels between 1996 and 2001 resulted in less than a 1% decline in farmland use.

The mix of crops farmed changed significantly in response to government policy changes, but Canadian cropland remained in production. Three crop groups historically account for just over half of Canada's total cropland: (1) wheat, (2) selected grains (oats, barley, and corn), and (3) selected oilseeds (principally canola but also including flaxseed, soybeans, sunflower, and mustard seed).

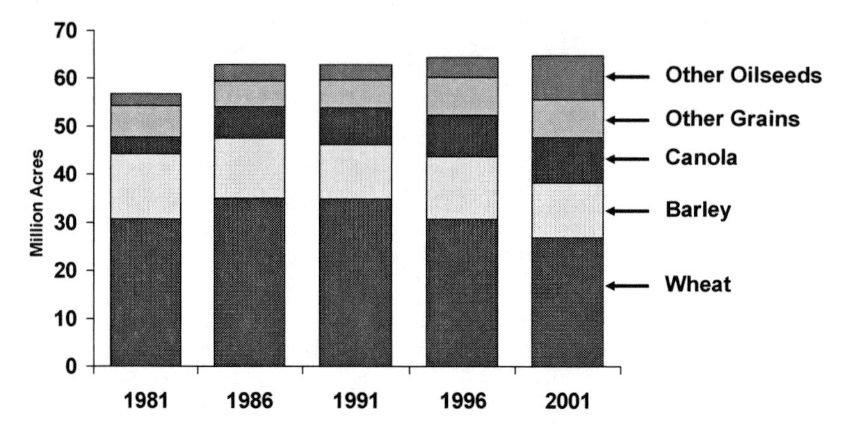

***Figure 3**. Canadian area planted to crops, 1981-2001*

Figure 3 shows the Canadian acreage planted to each of these three crop groups since 1981. Between 1991 and 2001, the acreage of Canada's leading crop, wheat, declined 23%. The elimination of subsidies for grain transportation in 1995 was a major contributor to this significant shift. Over the same period, oilseed production increased 143%. While the crop mix changed as relative prices and program payments changed, aggregate land in production changed little.

The Australian experience. The Australian experience also demonstrates the tendency of farmers to continue to produce as much as they can, even when faced with declining government subsidies. Since 1991, despite periods of low world prices, planted areas of wheat, coarse grains and oilseeds have increased more than 56% in Australia, as shown in Figure 4.

The Australian experience illustrates farmers' ability to shift resources from livestock to crop production in response to policy and price changes. Australia is the world's leading supplier of wool. Historically sheep and wool production has represented a large share of Australia's agricultural receipts. The Australian

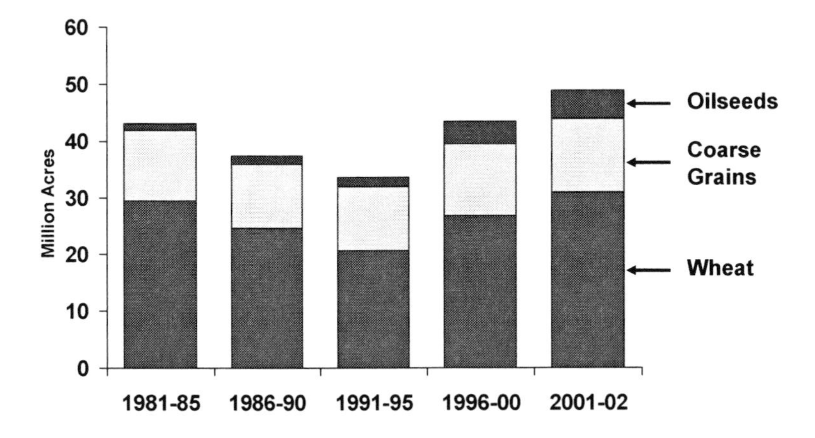

***Figure 4**. Australian area planted to crops, 1981-2002*

government's support for wool production was slashed in 1991. Since 1991, Australian sheep inventories have declined by over 30%. Faced with declining government supports for wool, sheep farmers converted significant pasture acreage to the production of wheat and other crops.

Are the WTO colours scrambled?

The preceding discussion gets us to these questions: What if conventional wisdom is dead wrong in the case of domestic price and income programs for agriculture? What if the worldwide elimination of all of the trade distorting price and income programs that have been identified by the WTO will not or cannot achieve the expected results?

What if, after the programs are eliminated, total production and the overall price level of major crops remain virtually unchanged from current levels – as the discussion in the preceding sections would suggest? What then?

As we have seen, it is the expanding size of agriculture's productive capacity that has the most depressing effect on prices. And yet, those public expenditures that expand productive capacity, including research and extension, general infrastructure and other capacity-building activities, are classified as non-trade distorting and put into the green box. Perhaps there is a need to re-examine the WTO classification system and the conventional wisdom that is attached to it.

Is it possible that all or most of the domestic agricultural programs are assigned to boxes of the wrong colour? If judged by the degree to which a domestic program depresses prices, an argument can be made that the blue box supply control programs and the amber box price support programs belong in the green box and the research, extension and many of the programs in the green box belong in the amber box. Of course, the box designation partially depends on the how each program is administered. If supply control and price support programs were used to raise prices well above the cost of production, the amber box comes back into the picture. If research, extension and other currently designated 'non trade distorting' activities are only invested to the extent required to maintain productive capacity and not to expand it, then such policies should logically remain in the green box. None of these possibilities seem likely.

The most striking conclusion of all this is that, given the mammoth and likely accelerating growth in productive capacity and the nature of agricultural markets, a subset of the domestic programs that the WTO and others condemn may be the very programs that are needed to prevent dumping and to achieve politically acceptable price levels, especially in developing countries. If those or other programs were accepted, most of the issues concerning government payments would be mute. First, as we have already established, government payments have little influence on total crop production. Second, payment levels would no longer be a major source of income if more price-oriented supply management policies were implemented. Third, US farmers and farmers worldwide would receive higher price prices and market incomes under supply management programs.

Unwise policy prescriptions

The 1996 change in US farm policy is a case of 'jumping from the frying pan into the fire'. Though never perfect, the historical application of the two major US commodity policy components – one expanding and the other throttling productive capacity – worked significantly better than the policies that were put in place with the adoption of the 1996 Farm Bill. One could argue that the traditional policy combination (1) was in the interest of food consumers, (2) facilitated orderly adjustments in farm scale and numbers, and (3) did not unduly disrupt farmers in developing nations by dumping excess production on the international markets. Indeed, the current agricultural policy configuration has made things worse in the eyes of farmers in developing countries and policy makers throughout the world by retaining policies to expand productive capacity and eliminating the ability to throttle that productive capacity. So what are the alternatives for US policy makers? There are a couple of alternatives that seem unwise or will not work.

It makes no sense to eliminate or vastly reduce public expenditures for research, extension, infrastructure and other activities that expand agriculture's productive capacity, even though the use of that capacity is one of the root sources of farm price and income problems. It is essential to keep well ahead of maximum demand needs not only for this generation but for generations to come. The best way to do that, given the relatively long cycle time for research, is to continue to invest in technologies that push productive capacity ever larger.

Another poor choice would be to completely eliminate *all* US commodity price and income support programs. Contrary to expectations, that approach would result in the same depressed prices and incomes for developing country farmers as the current government payment based US farm programs. Total agricultural output in the US would decline much less than most would expect. Yes, those US crops with relatively higher price and income supports compared to other crops would show a decline in production and world prices for those crops would increase somewhat. But, the crops that were substituted onto the vacated land would experience increased production and reduced prices. The net effect would be of little to no help to farmers in developing countries. US agriculture and rural communities, on the other hand, would be severely de-capitalized.

The biggest difference between the production costs of the US and Brazil is the land charge (Baumel et al. 2000). So, for example, if the US were to eliminate completely all commodity price and income support programs, the price of US farmland would fall until production costs were in line with those of other large producers like Brazil and Argentina. With lower land costs, US farmers would be able to sell their corn at $1.75 a bushel – a price no longer below the cost of production. If the price were to get to $1.75 that way, would farmers around the world benefit? US farmers could no longer be accused of dumping and yet farmers around the world would still be faced with the same low prices that plagued them in the years between 1997 and 2002.

Domestic price and support policies in the US and other developed countries that do not acknowledge and address the root problems are unlikely to fulfil policy objectives. Given that (1) expanding agricultural productive capacity continues to be deemed a worthwhile public endeavour, (2) consumers do not respond to lower prices by consuming more, and (3) crop farmers do not reduce their production in response to lower prices, the challenge is to find mechanisms that hold a portion of the productivity at bay until it is needed.

Alternate policy directions that take the unique characteristics of agriculture into account

The following sections present three sets of US policies that are designed to help overcome aggregate agriculture's relative inability to self-correct following even a significant decline in market prices. They could be viewed as short-run, intermediate-run and longer-term policy alternatives. The first set is based on the presumption that US markets largely determine world prices. This is, of course, more true for some commodities than for others. Events in recent years suggest that the US still plays a dominant role in international price determination. For example, the weather-shorten yields for major crops in 2002 and 2003 and the bumper crop in 2004 in the US substantially affected international crop prices. The second set merges farm policy and elements of energy policy. As a way to minimize the use of land set-asides to raise prices, a portion of that land could be used to produce energy crops. This policy component has the equally important benefit of reducing the world's dependence on fossil fuel. The third set involves the use of multinational cooperation to throttle international productive capacity to better approximate demand requirements at prices that cover production costs.

Each of these alternatives put a brake on output to help remedy aggregate agriculture's inability to make timely market corrections on its own. With higher prices, there would be less need for price-compensating government payments to US farmers. Prices also would be higher internationally and less 'excess' US production would be forced or dumped onto the international market.

Traditional policy instruments

The first set is a repackaging of policies that have been used in the US. It is a combination of (1) acreage diversion through short-term acreage set-asides and longer-term acreage reserves, (2) a farmer-owned food security reserve, and (3) price supports. The main objective of annual acreage set-asides is to avoid or to reduce the current tendency toward very low prices by inducing farmers to idle a portion of their working cropland. Land retirement in the form of a Conservation Reserve Program (CRP) – a tool already in use – serves to curb excess productive capacity in the long term. In identifying land to be put under CRP, farmers could select some of the most environmentally sensitive cropland. In this way they would ease the environmental burden caused by farming activities.

The second policy element, a food stock or inventory management reserve program, would reduce the occurrence and modify the size of major commodity price spikes. In exchange for a storage payment, farmers would enrol a share of their production in an on-farm storage program when prices are below a threshold level. When prices rise above the threshold, producers would be provided with an incentive to sell their reserves until the price dropped.

The third policy element, price supports, would provide an added measure to help avoid price collapses. Government price supports would be activated through government stock purchases triggered when prices fall below a threshold level. The government purchase of surplus stocks at a threshold level will eliminate the problem of dumping as long as the threshold level is set appropriately. This is in contrast to post-1996 policies, which have provided farmers with Loan Deficiency Payments (LDPs) and Marketing Loan Gains (MLGs) while allowing commodities to be sold on the world market at prices well below the threshold level.

If there is any one policy change that has made US agriculture vulnerable to the charge of dumping it is the institution of LDPs and MLGs – allowing the price to fall to so-called world levels – in the place of the traditional non-recourse loan program. The non-recourse program took low-priced grain off the market and did not allow it to flow into world markets, thus protecting farmers around the world from dumping by the US.

Our colleagues at the Agricultural Policy Analysis Center have used a simulation model to examine the impacts of this specific combination of policy measures on production levels and prices. The results of simulating these policy changes are remarkably clear: prices for the major commodities would increase from 23% for soybeans to more than 30% for corn, with rice and wheat not far behind. The general increase in the prices of all commodities would lead to net farm income levels close to and above that obtained through a continuation of the status quo, while at the same time reducing government payments significantly below the *status quo* projections, saving about $6 to $8 billion per year (Ray et. al. 2003, p. 46).

The $6 billion reduction in annual government outlays would certainly be good news for US taxpayers. And most importantly, perhaps, it would eliminate dumping US products onto into international markets. Because the US is the oligopoly price leader for many crops, higher US prices would be transmitted to the world market, helping to restore the prosperity for rural economies on which national economic development relies.

Intensifying agriculture's role in providing energy

Converting land into the production of dedicated bioenergy crops like switchgrass is one of the new and innovative means of addressing the need to manage the supply of storable crops. Instead of 'paying farmers not to farm' – an accusation made about acreage reduction programs in the past – a payment could be provided so that farmers would be able to provide the crop to a utility at a rate competitive with coal or bunker oil. As a perennial crop, switchgrass would help reduce soil erosion while remaining available for conversion back to crop production should the need arise. The payments could be directed in ways that strengthen farming as a livelihood strategy. The payments could also be targeted toward farmers who are within a certain radius of a co-fired electrical generation facility, leaving farmers at a greater distance to continue to grow their storable commodities.

In addition, switchgrass production could be targeted to areas facing serious disease or pest infestation by taking the land out of grain or seed production long enough to reduce the risk significantly. For instance, this could be important in nematode-infested fields for which a two-year corn–soybean rotation is not sufficient to reduce the nematode numbers.

With this approach, production levels could be managed by the diversion of acreage away from traditional tradable crops and toward the non-food, non-tradable crops such as switchgrass. When the annual set-aside was replaced with an incentive to develop a bioenergy-dedicated crop in the simulation model (retaining the farmer-owned food security reserve and price supports from the first alternative), results demonstrated overall levels of price increase comparable to those achieved by the set-aside policy. This illustrates that annual set-asides, while convenient, would not have to be a major component of the program.

Further, results similar to those demonstrated by introducing switchgrass could also be achieved by expanding the acreage enrolled in the Conservation Reserve Program (CRP). Such an approach may also contribute additional environmental benefits. Moreover, if necessary, land diverted to bioenergy-dedicated crops or placed in the CRP could be brought back into the production of major crops if unexpected weather or other events jeopardize the supply of food or if demand conditions warrant.

Multinational cooperation

Because the US is a major crop exporter and price leader, the previous two policy sets could be effective for the near future. But to sustain the improvement in farmer income over the long term, the US would have to be joined by other major agricultural players. A longer-run set of policies that would benefit farmers worldwide is the institution of an international program of supply management for both tropical crops like coffee and bananas and staples like corn, wheat, soybeans and rice. Such a program is needed because the major market characteristics that we have described for crops like soybeans and corn also apply to tropical crops. With the end of the Coffee Agreement and the investment by the World Bank and others to expand the geographic range of coffee production as a development tool, coffee prices have collapsed.

It is important to note that this price collapse happened in a crop that the US does not subsidize, because it does not produce it. US subsidies did not cause the sharp decline in coffee prices. Rather, the price problems that plague coffee, banana, cocoa, tea, jute and rubber reinforce our contention that the elimination of US and EU subsidies will not bring about the positive results that many expect. Once again, the key is to develop mechanisms by which agricultural production can be managed in a way that benefits both consumers and producers.

There again are three elements to this policy, but at a multinational level: (1) the establishment of an international humanitarian food reserve for essential storable crops and (2) the institution of a production management program by the top producers of a given crop, (3) coupled with a storage program to maintain prices within a predetermined range for storable crops. The international supply management program would be designed to benefit farmers worldwide. Domestic and international instruments need to be designed with the purpose of managing the use of the production capacity of agriculture in the countries of both the North and the South to the extent that countries are involved in export markets for those crops. Nothing is these agreements should prevent countries from enacting policies aimed at feeding their domestic populations. Only when a country began to export surplus production would it need to be a party to international supply management agreements. The establishment of these programs needs to be done within the context of the concept of food sovereignty.

Given a price goal or a reasonable price band, these instruments would allow producers a greater utilization of the production capacity during times of high prices and a lower utilization of the production capacity when prices are depressed. With the inclusion of a humanitarian reserve, areas of the world that experience random weather, pest or disease disasters would receive help. The storage program would help provide a band within which prices could vary. These programs would reduce the financial incentive to bring sensitive land like tropical forests into crop production by guarding against the risk of extremely high prices. In this and other ways, policies could be structured to achieve environmental, community, food sovereignty and other societal benefits.

SUMMARY AND CONCLUSIONS

Current US farm policy deserves much of blame for the depressed nature of world crop prices. But contrary to the usual arguments, excess crop production and fire-sale prices did not occur because farmers responded to payments and increased production. It occurred because the US no longer has the means to throttle its ever-expanding productive capacity or to establish a floor on commodity prices. Acreage set-asides and effective price supports are no longer part of the current US farm program, so all of agriculture's productive capacity is used all of the time.

The current US farm program does not work well because it is based on a model that treats agriculture as if it had the same characteristics as the typical manufacturing industry – it does not. The aggregate demand for food responds very little to wide swings in crop prices. Similarly, farmers do not take acreage out of production in response to, even, severe drops in price. For the markets to work in the same fashion as the typical manufacturing industry there would need to be price responsiveness on the part of supply and demand.

Recognizing this lack of price responsiveness, the US government has long enacted policies to establish a framework within which market forces could be brought to bear in the determination of crop prices. These policies took the shape of supply management programs that gave the US Secretary of Agriculture the tools to manage the total production of the US crop sector in much the same way that the CEO of Daimler-Chrysler determines the number of trucks and cars that will be produced in a given month.

Some general US policy alternatives that were identified are:

- Reinstitution of some of the traditional US farm policy instruments.
- Acreage reduction programs.
- Inventory management reserve.
- Establishment of a price band with a floor and ceiling.
- Intensify agriculture's role in providing energy.
- Gain multinational cooperation in an international supply management program.
- The following international implications were identified:
- Farmers in developing countries bore the brunt of the failure of the US to maintain its historic supply management program.
- A subset of the US supply and price support programs that the WTO and others tend to condemn may be what is needed to prevent dumping and to achieve politically acceptable price levels worldwide.
- The unique international characteristics of food and agriculture should be kept in mind when evaluating trade agreements as well as the impacts of changes in countries' agricultural policies.
- In light of the economic response and the food-security- and sovereignty-based political considerations that characterize food and agriculture worldwide, it is important to consider seriously whether the WTO colour classification system for agricultural programs and WTO's most fundamental premises with regard to the behaviour of the world's agricultural sector need to be re-evaluated.

NOTES

[1] Aggregate food consumption and per capita calorie intake will be affected somewhat with changes in the mix of food consumed as consumers respond to changes in relative prices of individual foods and

changes in tastes and preferences. Similarly, these forces may have a non-negligible affect on the intensity and mix of input use in agriculture and therefore a slight impact on total agricultural production, especially if changes in prices or in tastes and preferences shift consumption from low-input foods such as cereal to higher-input foods such as meat.

[2] In our view, it is not the purpose of US farm programs to lift farmers out of poverty or to allow all who want to farm to farm. Thus, if the market worked perfectly, there would be those who possess too few resources or lack sufficient management abilities to earn an 'adequate' living in agriculture. A portion of those may be eligible to receive help from other government programs.

[3] Not only does acreage change little with changes in prices, application of yield-determining inputs such as fertilizer, seed population and pesticides changes only slightly too. The fact that farmers can only influence their revenue by using adequate levels of the most productive inputs is not lost on US crop farmers or those that advise crop farmers. Any savings farmers achieve from reduced use of seed, fertilizer and pesticides tends to be small compared to revenue lost due to lower yields.

REFERENCES

Baumel, C.P., McVey, M.J. and Wisner, R., 2000. An assessment of Brazilian soybean production. *Doane's Agricultural Report,* 63 (25), 5-6.

Culver, J.C. and Hyde, J., 2001. *American dreamer: a life of Henry A. Wallace*. Norton, New York.

Halcrow, H.G., Spitze, R.G.F. and Allen-Smith, J.E., 1984. *Food and agricultural policy: economics and politics*. McGraw-Hill, New York.

Ray, D.E., De la Torre Ugarte, D.G. and Tiller, K.J., 2003. *Rethinking US agricultural policy: changing course to secure farmer livelihoods worldwide*. Agricultural Policy Analysis Centre, University of Tennessee, Knoxville. [http://apacweb.ag.utk.edu/blueprint/APAC%20Report%208-20-03%20WITH%20COVER.pdf]

Schaffer, H.D., 2004. *On predicting the price of corn, 1963-2002: selected paper prepared for presentation at the Southern Agricultural Economics Association Annual, Tulsa, Oklahoma, February 18, 2004*. Southern Agricultural Economics Association. [http://agecon.lib.umn.edu/cgi-bin/pdf_view.pl?paperid=12384&ftype=.pdf]

Schaffer, H.D., unpubl. *Econometric models on soybeans and cotton*.

Shover, J.L., 1965. *Cornbelt rebellion: the Farmers' Holiday Association*. University of Illinois Press, Urbana.

CHAPTER 13

THE WTO AGRICULTURAL NEGOTIATIONS AND THE LEAST DEVELOPED COUNTRIES

Limitations and options

SOPHIA MURPHY

Trade Program Director, Institute for Agriculture and Trade Policy, Minneapolis, Minnesota, USA

THE WTO NEGOTIATIONS ON AGRICULTURE: WHERE ARE WE NOW?

World Trade Organization (WTO) negotiations on the trade agenda agreed at the fourth WTO Ministerial Conference in Doha, in November 2001 – known as the Doha Agenda – have been paralysed for several years and finally ended in a state of deadlock in July 2006. After a failed Ministerial Conference in Cancún in September 2003, WTO members succeeded in taking their first step towards agreement on the Doha Agenda in 2004, with a document called the July Framework (also known as the July Package)[1]. The Framework did not represent much progress, considering the original deadline to complete negotiations on the Doha Agenda was January 2005, but it was the first agreement WTO members came to after Doha, and was widely acknowledged to have renewed impetus for the talks. The proposals are reviewed in more detail in section *"The July Package, Hong Kong, Next Stop Geneva?"*, below. The Ministerial Conference in Hong Kong, held in December 2005, made limited progress on most issues, and arguably least progress on agriculture. The Hong Kong outcomes are also reviewed below.

At the political level, negotiations on the July Framework reflected a new balance of power at the WTO. Particularly in the agriculture talks, the traditional USA/ EU hegemony is being challenged by Brazil, India and Australia. China, too, is playing an increasingly important role. Yet this new configuration of power, although it now includes the developing world, is still a long way from reflecting the needs and concerns of Least developed Countries (LDCs). Brazil, India and China

N. Koning and P. Pinstrup-Andersen (eds.), Agricultural Trade Liberalization and the Least Developed Countries, 233–249.

pose significant challenges to LDCs: Brazil is a source of cheap exports that displace local producers in LDC domestic markets, while all three compete with LDC exporters for access to the gradually liberalizing markets in OECD countries. China, in particular, is also a big buyer of some agricultural commodities, competing with LDCs for imports and able to outbid them. While facing considerable development challenges of their own, countries such as Brazil have far more resources with which to tackle development challenges than do LDCs such as Tanzania or Haiti.

Of course the expansion of the power base to include developing countries in agricultural negotiations is positive, undeniably so. Yet the more powerful actors in the agricultural trade negotiations still for the most part have other needs and concerns than LDCs. The bigger economies will ignore LDC needs when the full pressure of negotiations is on, as the significant limitations of the so-called 'development package' cobbled together in Hong Kong underline. It is still the dominant exporters – and to a more limited extent the largest importers – that control WTO negotiations. Despite their often significant dependence on trade in their own economies, LDCs are not part of either group.

LDCs AT THE WTO

Thirty-two of the 50 LDCs are now WTO members. Least developed Countries (LDCs), although identified as a distinct category in the Uruguay Round Agreement on Agriculture, have not worked very closely as a group in the negotiations on agriculture. LDCs are exempt from a number of disciplines under the Agreement on Agriculture (AoA), which contributes to some extent to their being ignored by other WTO Members. Few exporters see potential in LDC markets, and LDC production in most cases is too limited to create problems, or opportunities, in world market terms.

However, LDCs are not only affected because they have to implement WTO rules as member states. They are also affected because the other members' implementation of WTO rules changes the entire global trading context. For example, the liberalization of trade between the US and the EU has important implications for some LDCs. Perhaps even more important are the changing trade relations among developing counties and between developing and developed countries. For example, the erosion of preferences as the WTO ratchets down tariffs for all WTO members is diminishing a once protected market for some LDCs' exports, particularly to the European Union. Privileged market access becomes meaningless as tariffs get closer to zero.

The exemptions proposed for LDCs by other states, especially developed countries, create tensions within the larger group of developing countries. The EU proposal known as 'Everything but Arms,' to give LDCs duty-free access for most products, upset other vulnerable, non-LDC developing countries. Particularly for the Caribbean states, which are not LDCs but which currently enjoy preferential market access to the European Union as members of the Africa, Caribbean and Pacific (ACP) group, their export income is directly threatened by the extension of duty-free access to all LDCs. In any case, a number of developed countries, including Japan and the US continue to protect a number of products that would be of export interest to LDCs, including sugar, rice and cotton.

LDCs have great difficulty participating fully in WTO negotiations. Many LDCs lack representation in Geneva because they cannot afford to staff a mission. LDCs also have very limited staff available to focus on WTO issues in the national capital, making it difficult to ensure the negotiations reflect LDCs' national priorities. Many LDCs are hampered because they must manage a number of trade negotiations simultaneously. Those that are members of the ACP group of countries, for example, are in the midst of negotiating bilateral Economic Partnership Agreements (EPAs). Since the EU is the primary export destination for many of these countries, the EPA talks take priority over negotiations at the WTO. At the same time, many LDCs find their voices muted by pressure from bilateral donor countries, on which they depend for foreign aid money. LDCs, and developing countries that benefit from preference schemes, are also vulnerable to pressure by the preference-providing countries.

All of these problems are profoundly exacerbated by the *ad hoc* nature of WTO procedures and the lack of formal rules. The flexibility of the WTO way of conducting business is undoubtedly a factor to the WTO's success, which is why the richer countries are loath to change it. However, the flexibility comes at the expense of countries that cannot attend meetings at a day's notice or less, or which are not considered important enough to be invited to the informal session where compromise positions are hammered out. While LDCs are generally included in such processes as Mini-Ministerial conferences, it is generally a hand-picked few LDCs – just one or two – that are invited, while the others are left out.

Despite these difficulties, however, a number of LDC WTO Members have recently been active in two arenas. First, the African Group (which includes many LDCs) now meets to determine joint positions ahead of key negotiations, such as Ministerial Conferences and the negotiations this past July in Geneva. Second, the so-called Group of 90, which comprises WTO members from LDCs, ACP countries and the members of the African Union, has also met several times since its inception at Cancún. The G-90 agenda is fairly general – for agriculture, it includes the demand to end EU export subsidies and to increase market access to developed countries' markets. In Hong Kong, this group met with the Group of 20 (G-20), which is also a grouping of developing countries in the agriculture talks. G-20 members include Brazil, India and China.

A demonstration of the potential strength of the G-90 was evident in its refusal to negotiate on three of the four issues introduced at the first WTO Ministerial Conference, held in Singapore in 1996: investment, competition and government procurement. These issues were clearly rejected by the WTO's poorest members when they were proposed as components of the Doha Agenda, and the WTO members left Doha with only an ambivalent mandate for their possible inclusion. Ultimately, in Cancún, only the fourth Singapore issue, trade facilitation, was accepted for inclusion in the Doha Round. G-90 resistance to the inclusion of the other three issues was an important reason for their relegation to working group (as opposed to negotiating committee) status at the WTO.

A number of multilateral coordinating mechanisms, such as the African Union, the UN Economic Commission for Africa, the Commonwealth Secretariat, the ACP Secretariat and others have placed staff in Geneva in an attempt to provide LDC and other poor country WTO missions with more technical and political capacity. Nonetheless, LDCs cannot participate as full members of the WTO at the moment. The simple fact that the WTO holds more meetings simultaneously than LDCs have staff to cover them is a severe limitation on LDCs' ability to engage. The WTO's dispute settlement system is also out of reach, as it is too expensive for LDCs to contemplate bringing cases in their own right, despite the establishment of a legal-aid centre to provide advice and assistance on a more affordable basis.

DOHA TO CANCÚN

The Uruguay Round Agreement on Agriculture included a provision for its renewal in Article 20. Agriculture was thus always to be part of the post-Uruguay negotiating agenda. In Doha, at the fourth WTO Ministerial Conference held in November 2001, it seemed until the eleventh hour had come (and gone) that WTO members would reject a broad new agenda for negotiations. Developing countries resisted the proposed expansion of the agenda into a multifaceted single undertaking, arguing that as small delegations they already found it difficult to keep up with the pace of negotiations and fearing that they would be negotiating areas where their own interests were still unclear. However, into the night after the conference was due to close, the countries that sought a broader round (particularly the EU, Japan and South Korea) prevailed, aided by the countries with an ambitious agenda for agricultural liberalization – particularly the Cairns Group – which were convinced that significant new disciplines on agriculture would be impossible without the inclusion of other issues[2]. The Doha Agenda was born.

The Doha Ministerial Declaration was launched with much fanfare as a Development Agenda. The WTO website continues to use Doha Development Agenda as the title for materials related to the Doha Ministerial Declaration and the negotiations it gave rise to. The Doha Agenda set out an ambitious (if ambiguous) programme for agriculture. The language echoed that of the existing Agreement on Agriculture, but was influenced by some of the disappointments that marked the implementation of the Uruguay Round Agreement. The ambiguity came in the mention of export subsidies, where many members had hoped for a clear commitment to their elimination, but instead got something less precise: "reductions of, with a view to phasing out, all forms of export subsidies". The Doha mandate for agriculture was on the whole well received by WTO member states – all except some of the members of the European Union, which were not, at least not publicly, prepared to accept that export subsidies were doomed.

Doha raised important procedural and substantive concerns for developing countries, concerns that grew in Geneva in the months following Doha. For many developing countries, and especially the poorest members, a development agenda had to start by addressing problems with the Uruguay Round agreements: areas of ambiguity that needed clarification, rules that were having unintended and perverse consequences, and the failure to implement some of the flanking decisions that were intended to facilitate developing country engagement in the multilateral trading system. For example, the Marrakech Decision on Net-Food Importing and LDCs was never implemented, although increases in world prices for food in 1995 and 1996 would have warranted it. These issues are known collectively as implementation issues. One by one, the deadlines assigned in Doha to resolving implementation issues slipped. These were the issues that developing countries had insisted be included in the Doha Agenda, and which developed countries had used to justify their claim to meeting development needs in the negotiations. So, too, the deadlines to mark progress on the new negotiations passed unmet. Developed countries showed no willingness to compromise, despite their promises in Doha, and developing countries were holding fast to their positions as well.

On agriculture, points of division included:

- Interpretations of export subsidy language – to fix a date for elimination, or only to work towards eventual elimination.
- The level of ambition on market access – cut all tariffs by the same average amount, as had been done in the Uruguay Round Agreement, or create a formula that ensured the highest tariffs were cut by much more than already low tariffs, known as the Swiss formula. Harbinson proposed a banded approach, which would effectively combine elements of both approaches. In practice, this was a soft version of the Swiss formula.
- The scope and application of special and differential treatment – how generous should such provisions be, and should they be extended to all or only some developing countries? Should a new category, or some kind of graduated approach, be established for transition economies?
- The role of non-trade concerns and the legitimacy of developed-country appeals to be allowed to protect their 'food security' through Blue Box supports and high tariffs on particular commodities.
- The expansion of non-trade concerns – some countries want to include measures to allow higher standards to meet animal welfare concerns in livestock rearing, the use of the Precautionary Principle in setting food safety standards, and mandatory labelling (the last two concerns were immediately contentious because they were invoked in relation to conflicts between the US and the EU on the use and handling of genetically modified organisms).

As the date for the fifth Ministerial Conference approached, it became clear negotiators were working on the unspoken assumption that modalities would only be agreed to at the Ministerial Conference. This was a setback from the Doha timetable, but still on track for possible final agreement by January 2005. In August 2003, a month before Cancún, the US and the EU published a joint text on agriculture: a remarkable achievement, given their public differences on many issues. The political signal sent by the joint text, however, was out of step with the times. This was not 1992, when an agreement at Blair House between the US and the EU had been enough to clinch a deal. By now the WTO had 146 members, and a number of them had strong views on agriculture – particularly Brazil, India and China, but also groups such as the African Group; Least developed Countries (LDCs) and landlocked countries; ACP countries; and newly acceded countries, including economies in transition. The majority of WTO members were ready to fight rather than let the US and EU resume control of the agriculture talks.

The US/EU deal did not propose a date for the elimination of export subsidies, did not propose any limits on income support to farmers (an issue revisited below in the discussion of Green Box reform) and proposed that a category called 'sensitive products' be established for developed and developing countries alike that would allow higher levels of protection for some products, in the context of deep cuts to most tariffs. The reaction in Geneva was immediate, public and by diplomatic standards hostile: India and Brazil both rejected the deal as inadequate and self-serving. "This seems to be an attempt to pry open the developing country markets without any clear commitment on the part of (the US and EU) to open their own markets", Indian Ambassador K.M. Chandrasekhar told reporters. "I think it isn't feasible for us."[3]

More importantly, developing countries organized a written response. Brazil, India and China took the initiative to produce a counter-proposal a week later – still ahead of Cancún. They were initially joined by 13 other WTO members (some others joined later), thereby ensuring that the US/EU text could not serve as the *de facto* basis for negotiation in Cancún. The Group was the basis for what evolved into the Group of 20, or G-20[4]. Tanzania is the only LDC member of the group[5]. Although not historically allies on agriculture, Brazil, China and India were united in their determination to challenge the presumed leadership of the EU and the US. The G-20 text called for a clear date for the elimination of export subsidies, deep cuts to domestic support, and for distinct tariff formulae that would ensure deeper tariff cuts for developed countries than for developing. The differences among positions were deep and there was little time before Cancún to resolve them.

The fifth WTO Ministerial Conference was held in Cancún, Mexico, 10-14 September 2003. The meeting collapsed early, in failure. Only one decision was adopted: that the General Council would meet before December 15, "to take the action necessary at that stage to enable us to move towards a successful and timely conclusion of the negotiations"[6].

Although the Singapore issues were the immediate cause of the collapse of the 5th Ministerial Conference, it is widely agreed that agriculture was as great a problem. The first three days of the conference were largely spent on agriculture. The battle lines were drawn between the US and the EU (somewhat improbably united) on one side and the by-then Group of 21 (the membership fluctuated and has since settled at 20) on the other. A third grouping, the Group of 33 or G-33, also played its part, defending the position of developing countries that were less focused on increasing market access than the G-20, and more concerned with protecting rural livelihoods and food production at home. Recriminations flew, as the US and the EU blamed G-20 members (and Brazil in particular) for their failure to compromise, while many others pointed out the lack of real compromise from the two major players. It later emerged that the G-20 had prepared a compromise position, but the Ministerial Conference collapsed before it could be tabled.

Throughout the meeting, substantive progress was undermined by inadequate process. The text on agriculture sent by the General Council chairman, Perez del

Castillo, to Cancún, for example, largely ignored the contribution made by Brazil, China, India and the others, as well as other positions put forward by various groupings of developing countries in the months before Cancún. In Cancún, despite efforts to avoid the Green Room process of the GATT days, meetings continued to be held with restricted numbers of delegates. Perhaps more seriously, the WTO continues to operate in an entirely informal way, despite the formal – and legally binding – nature of its negotiated outcomes[7]. The lack of rules of procedure acts against smaller, weaker delegations, which are left without recourse when the informal process excludes them. This fuelled the longstanding sense of mistrust, particularly evident among the African and Caribbean countries at Cancún, and made a consensus outcome less likely. As the Third World Network reported a minister from the Caribbean saying: "What kind of organization is this? Who does it belong to? Who does the drafting? Who appointed them? Why waste our time engaging seriously in consultations only to find our views not there at all in the draft?"[8].

CANCÚN TO THE JULY PACKAGE

Sceptics doubted developing countries would maintain their Cancún alliances. Indeed – almost certainly under strong pressure from the United States – five Latin American countries (Colombia, Costa Rica, El Salvador, Guatemala and Peru) left the G-20 soon after the Cancún Ministerial Conference[9]. However, the groups have proved more resilient than expected. For example, a number of the G-20 members met in Argentina only one month after Cancún to reaffirm their shared purpose. Despite significant differences among the members, particularly on the question of market access, the G-20 has maintained a clear identity in the negotiations. The G-20 has now held two Ministerial Conferences, one in Brasilia in December 2003 and the second in São Paulo in June 2004. The countries meet as a group in Geneva on a regular basis and the Brazilian government hosts the group's official website.

The December 15-16 2003 meeting of the General Council focused on procedure rather than substance; informal meetings had made it clear there was still no agreement among members on the core areas of the Doha Agenda. However, the General Council did decide to restart meetings of the issue specific negotiating groups and the committee pulling it all together, called the Trade Negotiations Committee (TNC).

On January 11, 2004, US Trade Representative Robert Zoellick sent an open letter to all WTO-member Trade Ministers, to try to get negotiations moving again[10]. The letter proved to be highly influential on the content of the eventual July Package. Zoellick suggested negotiators focus on the 'core market access topics', meaning agriculture, services and industrial goods. Ironically for the countries and non-governmental organizations that had wanted to avoid a new round of comprehensive talks in Doha, Zoellick was now proposing something similarly restricted. The letter insisted that a clear end date for all export subsidies was a *sine qua non* for completing a new agreement on agriculture. In exchange, Zoellick offered to negotiate the elimination of the 'subsidy component' of export credits (a tool mostly used by the US).

His second challenge to other developed countries concerned domestic support: Zoellick proposed a cap on Blue Box spending. The US no longer uses the Blue Box, but many other OECD members use it extensively. For the United States, Zoellick made it clear that two things would determine the level of cuts to domestic support they could agree to. One, countries with relatively higher AMS and Blue Box spending would have to make the deepest cuts (the cuts should be proportionate to spending levels, not identical across the board) and two, the level of cuts to domestic support would have to be matched by real and significant market access to both developed and developing countries.

The letter also addressed SDT. The letter clearly said progress on agreeing meaningful SDT provisions would depend on establishing criteria that narrowed the list of countries eligible for special treatment. Zoellick conceded the request by the G-33 to create a category of 'special products,' linked to concerns about rural development and subsistence farmers, who would be allowed higher levels of protection – but only for a 'very limited number' of crops. On the Singapore issues, Zoellick put forward the EU's compromise offer from Cancún (which the Commission had subsequently withdrawn) to 'unbundle' the issues and consider taking up one or more, but not all four, in the context of the Doha negotiations.

Finally, Zoellick called for a mid-year meeting to decide 'frameworks' for the negotiations, and for the next Ministerial Conference to be held ahead of schedule, by the end of 2004. The suggestion of moving the Ministerial to 2004 was quickly dismissed as unrealistic, but the idea of a mid-July high-level meeting in Geneva to agree frameworks did gain traction. The value of the letter was as much psychological as practical. It gave renewed impetus to the talks, showing the US was prepared to put political energy into the WTO.

Another important development over the first half of 2004 was European Union progress on Common Agricultural Policy (CAP) reform. In June 2003, the European Commission announced fundamental reforms for the CAP. Most significantly, payments were by and large to shift from payments linked to production to decoupled payments. The reforms freed the European Commission to play a more active role in the Doha negotiations because the CAP reforms are bringing most CAP spending into compliance with AoA disciplines.

THE JULY PACKAGE, HONG KONG, DEADLOCK IN GENEVA

Shortly after midnight on the 1st of August 2004, members of the WTO agreed a framework that provided a basis to continue negotiations on the Doha Agenda. A short overview text is accompanied by a series of annexes; Annex A deals with agriculture. Only one of the Singapore issues – trade facilitation – survived as a topic for negotiation, although the other three continue to be topics of discussion in working groups. Overall, the July Package reflects and expands on the Doha Agenda, with one or two changes and additions, together with the addition of considerable detail. A great deal remains to be negotiated, however, and the most recent Ministerial Conference, held in Hong Kong in December 2005, suggests WTO members are still very far apart on many areas under negotiation. In Hong Kong, WTO members gave themselves a deadline of April 30, 2006 to agree another series of difficult points, to keep the negotiations in motion. (The decisions actually taken in Hong Kong, particularly as regards agriculture, were very few, as is reviewed below). At the end of June, ministers and heads of delegations met in Geneva to negotiate once more on "modalities". However, the negotiations were suspended after an attempt by ministers from six key players to break the deadlock failed on 23 July. Final agreement on the Doha Agenda continues to look unlikely any time soon.

The July 2004 Framework reflected many of the elements that Zoellick proposed in his January letter: export subsidies will be eliminated (although not any time soon); the Blue Box will be capped (although also expanded to include programs that are not aimed at limiting production); and market access will be increased on all products (although some products will be granted less stringent tariff reductions than the rest). Some SDT measures are mentioned, but no detail is given. LDCs continue to be exempt from many provisions.

Up close, the proposals in the July Framework do not add up to much. Given the Doha Agenda's stated aim for agriculture: "to establish a fair and market-oriented agricultural trading system", progress on reducing government payments to the agricultural sector was limited. WTO members agreed to eliminate export subsidies – an outcome that received very favourable coverage – but the end-date was not set (finally, in December 2006, the E.C. limped to a commitment, promising to eliminate export subsidies but only by 2013, and only if several other forms of export support are also disciplined, including food aid, state trading export enterprises and export credits). Developed countries agreed to tighten the disciplines on their spending to support domestic agriculture, but not by much. All of the detail on how much tariffs will be cut and on which products was left open. So was the detail on what special and differential treatment for developing countries, including LDCs, would actually consist of.

Aside from the date for the elimination on export subsidies, and some clearer commitment from all members for the special products and special safeguard mechanism sought by developing countries for their agriculture, the Ministerial conference in Hong Kong added little to the agenda. Some progress on bands for tariff and domestic support reduction has been made since July 2004, and was confirmed in Hong Kong, but the detailed proposals from the E.C. and the US (presented in October 2005) revealed that little significant change to agriculture policy will be realized in the new agreement[11]. Not least, there is simply no support in their respective Parliaments for dramatic liberalization of agriculture that departs from the ongoing domestic reforms now in place. Even in the case of cotton, where the US was found by the WTO dispute body to be in violation of the Uruguay Round commitments to refrain from introducing new export subsidies and to ensure farm payments respected 'least trade distorting' criteria, the US is unable to offer four of the poorest WTO members even a gesture of political goodwill. Ambiguous statements about reducing domestic support to cotton ahead of other commodities were the extent of the undertaking in Hong Kong.

CHALLENGING THE ASSUMPTIONS OF MARKET LIBERALIZATION

Governments at the WTO identify three primary sources of distortion in world agricultural markets: export subsidies, domestic support, and market access barriers. The AoA-mandated reductions to these distortions did not, in most cases, change existing spending or increase market access in any significant degree. However, the categorization of programs was in some ways more significant than the spending limits set. The categorization was important because it sent a signal as to what kinds of programs would be acceptable in the future, and pressured WTO member states to shape their agricultural programs in a particular way. In practice, the agreement discouraged payments to producers that were aimed at limiting output. The Blue Box (article 6.5 of the AoA) created an exemption for production-limiting programs but could only be based on *historical*, not actual production, making the exemption of limited use for governments intent on managing changes in supply and demand needs in an on-going way. So-called decoupled payments, which are based on historical rather than actual production, were blessed as 'non trade distorting' although they have been demonstrated to encourage production. At the same time, the AoA put a ceiling on tariffs, while many developing countries had already lowered their tariffs unilaterally to meet their obligations under structural adjustment programs. The AoA narrows the options available to countries to determine what kind of agricultural development model they want to pursue.

Were the WTO negotiations in agriculture to succeed in eliminating all government-related sources of market distortions in world agricultural markets, would LDCs be better off? Even though such a clear outcome is rendered all but impossible by the politics of agriculture in developed countries, it is also important to ask whether the WTO is pushing its membership towards rules that make sense. In fact, assuming that perfectly open markets in agriculture were politically possible, the nature of agricultural markets themselves would then create another set of challenges for policy makers. Left without regulation, agricultural markets tend to over-production, unsustainable production, and price depression, interrupted by periodic (and, for consumers, devastating) price spikes. Market power in commodity production and processing accumulates in the centre, creating oligopolies that are difficult to regulate and yet which are known to raise prices and diminish welfare. Here are some facts that challenge the assumptions that underpin the WTO negotiations on agriculture.

1. Many sectors have welfare implications, but more is at stake when it comes to agriculture. Unemployment is a cruel hardship, but starvation is fatal. Governments have an obligation to protect food security, and therefore must step in when food markets fail, and may need to regulate markets to protect people's access to food, even if that means distorting the market. Food security is protected in international law. UN member states are bound to protect and promote the universal human right to food. In an immediate sense, this means governments are bound not to restrict the policy space they need to move towards implementation of the right to food. Governments are also bound by their commitment to ensure food security, defined at the World Food Summit in 1996 as: "Food that is available at all times, that all persons have means of

access to it, that it is nutritionally adequate in terms of quantity, quality and variety, and that it is acceptable within the given culture"[12]. Governments cannot cut food imports the way they might decide to do without cars, or even fuel, if they had to.

2. The model for agricultural trade implicit in the AoA presumes that agricultural markets are only distorted by government interference. In practice, as researchers have increasingly documented, agricultural markets are heavily distorted by the presence of oligopoly buyers and sellers at different points in the agri-food chain. Monsanto, for example, makes some 90% of the genetically modified seed in commercial use. Cargill, ADM and Zen Noh control over 80% of maize sales from the US (which in turn has some 40% of the world maize market). Cargill, ADM and Bunge overwhelmingly dominate soybean sales from three major exporters: the US, Argentina and Brazil. Food retailers are also increasingly concentrated, and are penetrating developing country markets. Meanwhile, the countries most dependent on primary commodity exports – many of them LDCs – have been losing global market share over the past twenty years.
3. The external effects of farm production (the impact agriculture has on the environment, for example) pose another set of challenges to the assumption that agriculture can be left to the free market. For example, research in the Philippines has shown that while green-revolution technologies produce more rice per hectare than traditional methods, overall productivity on farms using green-revolution inputs is lower because the pesticide-filled water in the paddy no longer supports fish stocks[13]. These fish provided an additional source of protein for the household, which was a benefit not captured in the market equation, but which was very important to the well-being of the farm household.
4. Agricultural markets are distinct from other markets. Demand is both slow to respond to changes in price and unable to respond much, even over time. There is only so much food a body can digest, even when the food is free; new demand must come from alternative uses for the crops – making liquor, plastic or fuel additives – or from rising incomes, which increases the demand for value added crops (but is not a short-term solution to over-supply and falling prices). Supply responses are also slow and then often excessive (especially when prices spike upwards). Millions of individual farmers cannot affect price by changing their output; they are price-takers in commodity markets. Agricultural assets are fixed, making it hard to respond to price increases with increased supply, while the information available to individual farmers is imperfect, again inhibiting the ideal open market response. Poor information contributes to farmers over-reacting to price increases, leading to a new glut and further price depression in a cycle described as a cobweb, as each circle takes you further from the (ideal) centre. Agricultural supply is largely dependent on the weather, which is not an element farmers can control, although technology helps to manage its unpredictability. Together, these characteristics contribute to price volatility, where brief price spikes are followed by longer-lasting price declines. Futures and options markets, which are often proposed as a tool to reduce price risks, fuel unstable prices, particularly, as with grains and oilseeds, when the spot market is small relative to contracted production. Agricultural markets have

distinct characteristics from those that characterize the ideal in the dominant free-market model.

5. The second half of the 20th century presented the world with a new problem: what to do with surplus production, not just from a good year but also from massive increases in productivity year after year. At the turn of the 20th century, over 25% of agricultural production was used to feed the animals that helped till the soil and move people and goods. Then the combustion engine took over. Most of this land is now dedicated to growing food, adding to the surpluses. We need effective tools to manage production, not least to limit the damage caused by the technologies the productivity relies on. But we also need to keep an eye to the future, and plan for long-term food security, not just for keeping costs down today. We may not want any more yellow maize grown in the US (in fact, we may want a lot less), but we may want to keep the land used to grow that corn available for agricultural production in the future – either for food or for fuel. World population growth has moderated but the world's population is still increasing, while the total arable land available to the planet is not.
6. There is strong empirical evidence to show agricultural development is an effective – perhaps the most effective – way to generate employment and reduce poverty in developing countries. Increasing incomes in rural areas has an immediate and significant positive effect by increasing demand for local goods and services. It is those living without land in rural areas – the people who generally provide these services – who make up the majority of the extremely poor. Their livelihoods should be the first concern of the international community.
7. Climate change is posing many new challenges for agricultural planning. Climate change already affects agricultural production and most experts agree that it will have much more effect in the near future. Experts forecast that valuable agricultural land in Bangladesh and elsewhere will disappear under water as the world's glaciers melt. Although warmer climates in some places may increase their agricultural potential, the most likely change seems to be increased instability in weather patterns. Trade is likely to be an invaluable tool for societies and economies when they tackle these problems, but current trade patterns, to some extent locked in by WTO rules, inhibit needed initiatives to develop alternative energy uses and to transform resource use and management.

Despite the failures of the model in both theory and practice, many governments continue to assert that WTO rules for agriculture reflect a balance of interests among exporters and importers. However, a review of the rules and their implications suggests that a rather narrower set of interests is actually coming out ahead. In particular, the interests of the transnational agribusiness firms that most actively engage in world markets are strongly represented by governments of both developed and developing countries. Meanwhile, the WTO itself clearly equates increased trade with increased human welfare despite the clear debunking of such simplistic assertions by numerous development analysts. The culture and the working methodology of the WTO favour exporters over importers. Exporters are the *demandeurs*, and countries whose agriculture is predominantly for export have less to lose in structuring their economies to favour increased trade. Yet these countries are a minority of the world's countries: 27 or so countries are net food exporters. And an estimated 85% or more of world agricultural production never crosses an international border.

Countries face complicated trade-offs in assessing the best approach to tariff, domestic support and export strategies. Competing interests and scarce resources mean policy choices are highly contested. Some traders see the WTO as a place to pass multilateral rules that could never be agreed to domestically. Analysts argue that this was part of the motivation for the Reagan administration in its push to secure an Agreement on Agriculture as a part of the Uruguay Round. Yet if this was the strategy, it has had unintended consequences: in many countries the public does not see the WTO as a legitimate actor because there is little support for the rules imposed under the Uruguay Round agreements. Many farm organizations, together with trade unions, environmentalists, Church-based groups and others oppose many aspects of the WTO (and some groups reject it outright). Few of the these groups, and they include some parliamentarians, trust the organization to deliver rules for multilateral trade that take account of development needs, human rights, livelihood issues and other public policy priorities.

WHAT CAN BE DONE?

Agricultural commodity markets are plagued by over-production and depressed prices. Some consumers have benefited from lower prices but many of the world's poorest consumers depend on higher commodity prices for their welfare. Increasing levels of concentration in global commodity markets undermine the effectiveness of price transmission. The following five proposals are offered in light of this analysis.

1. *Stronger rules against dumping.* Current WTO rules tackle dumping by allowing countries to tax imports that are sold for less than the price in the home market, but this levy is insufficient. Extensive and chronic over-production of many commodities have depressed prices and made dumping endemic. Dumping should be measured against production costs including a normal return to labour and capital, not against heavily manipulated domestic prices. Developing countries, unable to protect their producers with subsidies, must be allowed to block dumped imports immediately at the border[14].
2. *Stabilize commodity prices.* Agricultural commodity markets are inherently unstable (e.g. due to crop failures) and prone to both price spikes and prolonged periods of over-production and low prices. Unregulated commodity markets have failed to manage these structural characteristics (consider the recent roller-coaster ride of coffee prices, which has devastated the lives of millions of poor coffee growers). The UN Commission on Trade and Development has established a task force to create a toolbox of policies to address the global commodity crisis, including international commodity agreements among major exporters, and regional grain reserves. These policies must be implemented.
3. *Regulate market concentration.* Vertical and horizontal concentration in global commodity markets is a primary cause of market distortion. Possible policy responses include an international review mechanism for proposed mergers and

acquisitions among agribusiness companies that are present in a number of countries simultaneously. At a minimum, transparency requirements now imposed on state trading enterprises should be extended to companies with 20% or more of a national or global market in any given commodity.

4. *Link tariffs to supply management and export controls.* The 1947 General Agreement on Tariffs and Trade (GATT) allowed countries to use import quotas on agricultural products if they practiced supply management and did not export surpluses. This approach should be revisited. Instead of assessing national programs by how much they cost, trade negotiators should focus on their trade-distorting impact.
5. *Protect standards and national development.* As the WTO itself has observed, lowering trade and investment barriers makes regulation of industry more difficult, creating a trade-off between increased efficiency and strong standards, whether environmental, labour-related or other. Governments should approach competition and investment issues with a view to protecting standards and national development objectives. The current approach instead emphasizes the obligation on governments to demonstrate that regulations are 'least trade restrictive'. All WTO member states are obligated under international law to uphold and implement the Universal Declaration on Human Rights and at least some of the related Covenants. International trade law should not take precedence over commitments to protect human welfare.

CONCLUSION: A MULTILATERAL TRADE SYSTEM FOR LDCS

LDCs have a lot to gain from a transparent and predictable global trading system, where large countries are discouraged from cheating by a system of enforceable rules, and where collective bargaining is possible, uniting poorer countries with shared interests. The market distortions that plague developed country agriculture must be stopped. This argues for LDC membership in the WTO. Nonetheless, if LDCs were in charge, the WTO negotiating agenda would look quite different.

A trade agenda for LDC agriculture would contribute to a more productive and profitable agricultural sector. Employment creation and sustainable management of often scarce and depleted natural resources are of the highest priority. LDCs are not big players in global markets, but are unusually dependent on trade for their income. This includes both foreign exchange earnings from the sale of exports and tariff revenue from imports. As trade policy expands to include services (and therefore investment), LDCs bring a distinct set of concerns to the table: they need foreign capital and know-how, but directed at building their own firms and creating domestic employment and capital, rather than a system in which profits are taken out of the country and ownership does not transfer to local control. Such infrastructure is needed for local and regional markets, not just to facilitate access to world markets.

To date, the liberalization of agriculture (for example, the end of price floor policies in the US and of efforts to coordinate supply policies through international commodity agreements) has increased price volatility in commodity markets. Although long-term price declines continue, spot prices are volatile and food import bills vary significantly from year to year for LDCs, especially when currency fluctuations are factored in. The strong pressure to reduce tariffs further in the Doha round will further lessen the value of the preferences on which some LDCs depend. LDCs tend to be less competitive than other developing country exporters, and experience suggests that in a deregulated, open market context, they will find it hard to catch up. All this creates problems that the WTO, as a system of one-size-fits-all rules that deliberately seeks to avoid concessions to non-trade concerns where possible, is ill-equipped to address.

The interest of LDCs in the WTO as an institution is also worth reviewing. Smaller, weaker countries look to the multilateral system for strength in numbers and protection from unilateral action or decisions that exclude them. However, the WTO is perhaps the multilateral forum that finds it hardest to fulfil this promise. There is unquestionably a trade-off between the WTO's strength and the WTO's ability to handle a broadly-based development agenda (it has refused to do this to date). The strength comes from the WTO's clear – and narrow – focus, the support and interest of the world's largest and most powerful economies (which give the institution more attention and proportionally more resources than many of its UN cousins) and the continuing vogue for trade liberalization as the pre-eminent engine for economic growth, although nay-sayers are there for those listening.

There is empirical evidence to show that the relationship between trade liberalization and growth, particularly growth that generates employment and reduces poverty, is a complicated one. The WTO has the potential to provide a forum for weaker voices; it is a multilateral organization that formally gives each member an equal voice (unlike, say, the World Bank). However, in practice, it is widely acknowledged that smaller economies, and LDCs in particular, are not able to participate on equal terms with larger countries. The most recent Ministerial Conference in Hong Kong showed again that the relatively informal and *ad hoc* negotiating system works against the participation of many countries, including the majority of LDCs. LDCs often have to depend on one or two voices in the Heads of Delegation meetings. Reports of unfair pressure by other WTO members, as bilateral donors or preference-offering countries also persist, contributing to a climate in which developing countries are suspicious and feel marginalized.

Many LDCs have come through extensive economic structural adjustment programs and are now operating in what UNCTAD terms a post-liberalization context. Their challenges are not so much in how to open their markets further, or in how to manage greater liberalization, but rather in how to make up for not having been able to sequence policy changes properly or to invest in the necessary supportive policies to ensure that the potential gains from liberalization were realized. In that context, the multilateral trade rules matter a great deal, but they cannot hope to be the main engine of growth and development for LDCs. Membership of the WTO is probably useful for most developing countries, but it cannot substitute for investment in diversification of the economy, meeting supply-side constraints and developing options for those who have lost their livelihoods in the process of moving to a market-based economy.

Many farmer and peasant associations in both the developed and the developing world, together with the non-governmental organizations (NGOs) they work with, question the legitimacy of the WTO to determine domestic agricultural policy. Many of these groups have joined a campaign that calls for the 'WTO out of agriculture' on the grounds that the attempt to shape domestic agricultural policies through an international trade law lens is bad for farmers and undermines the basis for a just and sustainable food system. Increasingly, a convergence among NGOs is evident, where, despite differences on what value to place on more liberalized trade as an engine for growth, many NGOs agree that developing countries need to maintain national policy flexibility to determine the best policy mix for their needs. These organizations are interested in strong multilateral rules for agricultural trade. They accept that many problems can only be solved by multilateral disciplines. However, they are sceptical that the WTO's pursuit of deregulated markets responds to developing country needs.

For their part, many developing country governments are still hopeful that the WTO can be made responsive to development needs. They are trying to use the Doha Round negotiations on agriculture to make the rules more responsive to their needs. Many developing countries are now pursuing a strategy to secure an outcome that will actually liberalize developed country agriculture while seeking to protect some of their vulnerable agricultural populations. LDCs are not adequately represented in the debates, but many have associated themselves with the wider developing country efforts. At the same time, LDC governments have joined efforts such as the G-90 to attempt to consolidate the voice of the poorest WTO members. This paper argues they should go much further in their proposals, to build a protected space within which to secure the best development outcome for people, using agriculture as an engine for growth.

It is questionable at this point if scarce LDC trade capacities are well spent on the WTO. Certainly those that are not yet members could ask themselves if the costs entailed are worth paying. At this point, WTO agreements curtail policy space, reflect interests of dominant member countries, while the accession process imposes even greater liberalization conditions than the WTO Uruguay Round Agreements require. Those LDCs that are WTO members might challenge the assumption that proceeding on the basis of one-size-fits-all rules in a multi-sectoral take it or leave it package is appropriate to their development needs. The WTO has yet to prove itself as able to meet the demands of a diverse and, in too many cases, deeply impoverished membership.

NOTES

[1] The texts of the 'July Package' are at the WTO website: http://www.wto.org/english/tratop_e/dda_e/draft_text_gc_dg_31july04_e.htm

[2] The Cairns Group of agricultural exporting countries at present has 17 members: Argentina, Australia, Bolivia, Brazil, Canada, Chile, Colombia, Costa Rica, Guatemala, Indonesia, Malaysia, New Zealand, Paraguay, the Philippines, South Africa, Thailand and Uruguay.

[3] 'Cool Reception for EU, US Agreement on Framework for Farm Subsidies Cuts in WTO Talks' story by Naomi Koppel, Associated Press. 14 August, 2003.

[4] The G-20 now has a website at http://www.G-20.mre.gov.br.

[5] There are currently 19 members of the G-20: from Africa – Egypt, Nigeria, South Africa, Tanzania and Zimbabwe; from Asia – China, India, Indonesia, Pakistan, Philippines and Thailand; and from Latin America – Argentina, Bolivia, Brazil, Chile, Cuba, Mexico, Paraguay and Venezuela.

[6] As cited in *Bridges*, No.7, September-October 2003, International Centre for Trade and Sustainable Development. Geneva. On-line at www.ictsd.org.

[7] Amrita Narlikar, 'The World Trade Organization: A Case for G-20 Action on Institutional Reform,' conference paper *Agricultural Subsidies and the WTO*, Oxford, UK, June 8-9, 2004. On-line at: http://www.globalcentres.org/html/project9.html.

[8] Third World Network, 'Analysis of the Collapse of the Cancún Ministerial,' TWN Info Service on WTO Issues (Sept 03/14), 16 September 2003. On-line at http://www.twnside.org.sg/title/twninfo76.htm.

[9] There are now 19 members in the Group – see footnote 5 for the current list.

[10] Available on-line at a number of sites, including www.tradeobservatory.org in the library and at www.ictsd.org. The USTR's office has a summary version at http://www.ustr.gov/Document_Library/Press_Releases/2004/February/Zoellick_Embarks_on_Global_Push_to_Make_Strong_Progress_on_Doha_Negotiations.html.

[11] See for example the analysis by IATP, *The United States WTO Agriculture Proposal*, October 2005. On-line at http://www.tradeobservatory.org/library.cfm?refID=77195.

[12] FAO, 1996, Rome Declaration on World Food Security.

[13] Yap (Masipag), personal communication.

[14] Sophia Murphy, Ben Lilliston and Mary Beth Lake (2004), *WTO Agreement on Agriculture: A Decade of Dumping*, IATP. Minneapolis. USA.

Wageningen UR Frontis Series

1. A.G.J. Velthuis, L.J. Unnevehr, H. Hogeveen and R.B.M. Huirne (eds.): *New Approaches to Food-Safety Economics*. 2003
ISBN 1-4020-1425-2; Pb: 1-4020-1426-0
2. W. Takken and T.W. Scott (eds.): *Ecological Aspects for Application of Genetically Modified Mosquitoes*. 2003
ISBN 1-4020-1584-4; Pb: 1-4020-1585-2
3. M.A.J.S. van Boekel, A. Stein and A.H.C. van Bruggen (eds.): *Proceedings of the Frontis workshop on Bayesian Statistics and quality modelling*. 2003
ISBN 1-4020-1916-5
4. R.H.G. Jongman (ed.): *The New Dimensions of the European Landscape*. 2004
ISBN 1-4020-2909-8; Pb: 1-4020-2910-1
5. M.J.J.A.A. Korthals and R.J.Bogers (eds.): *Ethics for Life Scientists*. 2004
ISBN 1-4020-3178-5; Pb: 1-4020-3179-3
6. R.A. Feddes, G.H.de Rooij and J.C. van Dam (eds.): *Unsaturated-zone modeling*. Progress, challenges and applications. 2004 ISBN 1-4020-2919-5
7. J.H.H. Wesseler (ed.): *Environmental Costs and Benefits of Transgenic Crops*. 2005 ISBN 1-4020-3247-1; Pb: 1-4020-3248-X
8. R.S. Schrijver and G. Koch (eds.): *Avian Influenza*. Prevention and Control. 2005
ISBN 1-4020-3439-3; Pb: 1-4020-3440-7
9. W. Takken, P. Martens and R.J. Bogers (eds.): *Environmental Change and Malaria Risk*. Global and Local Implications. 2005
ISBN 1-4020-3927-1; Pb: 1-4020-3928-X
10. L.J.E.J. Gilissen, H.J. Wichers, H.F.J. Savelkoul and R.J. Bogers, (eds.): *Allergy Matters*. New Approaches to Allergy Prevention and Management. 2005
ISBN 1-4020-3895-X; Pb: 1-4020-3896-8
11. B.G.J. Knols and C. Louis (eds.): *Briding Laboratory and Field Research for Genetic Control of Disease Vectors*. 2006
ISBN 1-4020-3800-3; Pb: 1-4020-3799-6
12. B. Tress, G. Tress, G. Fry and P. Opdam (eds.): *From Landscape Research to Landscape Planning*. Aspects of Integration, Education and Application. 2006
ISBN 1-4020-3979-4; Pb: 1-4020-3978-6
13. J. Hassink and M. van Dijk (eds.): *Farming for Health*. Green-Care Farming Across Europe and the United States of America. 2006
ISBN 1-4020-4540-9; Pb: 1-4020-4541-7
14. R. Ruben, M. Slingerland and H. Nijhoff (eds.): *The Agro-Food Chains and Networks for Development*. 2006 ISBN 1-4020-4592-1; Pb: 1-4020-4600-6
15. C.J.M. Ondersteijn, J.H.M. Wijnands, R.B.M. Huiren and O. van Kooten (eds.): *Quantifying the Agri-Food Supply Chain*.
ISBN 1-4020-4692-8; Pb: 1-4020-4693-6
16. M. Dicke and W. Takken (eds.): *Chemical Ecology*. From Gene to Ecosystem. 2006 ISBN 1-4020-4783-5; Pb: 1-4020-4792-4
17. R.J. Bogers, L.E. Craker and D. Lange (eds.): *Medicinal and Aromatic Plants*. Agricultural, Commercial, Ecological, Legal, Pharmacological, and Social Aspects. 2006 ISBN 1-4020-5447-5; Pb: 1-4020-5448-3

Wageningen UR Frontis Series

18. A. Elgersma, J. Dijkstra and S. Tamminga (eds.): *Fresh Herbage for Dairy Cattle.* The Key to a Sustainable Food Chain. 2006
    ISBN 1-4020-5450-5; Pb: 1-4020-5451-3
19. N. Koning and P. Pinstrup-Andersen (eds.): *Agricultural Trade Liberalization and the Least Developed Countries*. 2007
    ISBN 978-1-4020-6079-3; Pb: 978-1-4020-6085-4
20. A.G.J.M. Oude Lansink (ed.): *New Approaches to the Economics of Plant Health.* 2007 ISBN 1-4020-5825-X; Pb: 1-4020-5826-8
21. J.H.J. Spiertz, P.C. Struik and H.H. van Laar (eds.): *Scale and Complexity in Plant Systems Research.* Gene-Plant-Crop Relations. 2007
    ISBN 1-4020-5904-3; Pb: 1-4020-5905-1
22. J. Vos, L.F.M. Marcelis, P.H.B. de Visser, P.C. Struik and J.B. Evers (eds.): *Functional-Structural Plant Modelling in Crop Production.* 2007
    ISBN 978-1-4020-6032-8; Pb: 978-1-4020-6033-5